Pro Machine Learning Algorithms

A Hands-On Approach to Implementing Algorithms in Python and R

D0882327

V Kishore Ayyadevara

Apress®

Pro Machine Learning Algorithms

V Kishore Ayyadevara
Hyderabad, Andhra Pradesh, India

ISBN-13 (pbk): 978-1-4842-3563-8   ISBN-13 (electronic): 978-1-4842-3564-5
https://doi.org/10.1007/978-1-4842-3564-5

Library of Congress Control Number: 2018947188

Managing Director, Apress Media LLC: Welmoed Spahr
Acquisitions Editor: Celestine John Suresh
Development Editor: Matthew Moodie
Coordinating Editor: Divya Modi

Cover designed by eStudioCalamar

Cover image designed by Freepik (www.freepik.com)

Distributed to the book trade worldwide by Springer Science+Business Media New York, 233 Spring Street, 6th Floor, New York, NY 10013. Phone 1-800-SPRINGER, fax (201) 348-4505, e-mail orders-ny@ springer-sbm.com, or visit www.springeronline.com. Apress Media, LLC is a California LLC and the sole member (owner) is Springer Science + Business Media Finance Inc (SSBM Finance Inc). SSBM Finance Inc is a **Delaware** corporation.

For information on translations, please e-mail rights@apress.com, or visit http://www.apress.com/ rights-permissions.

Apress titles may be purchased in bulk for academic, corporate, or promotional use. eBook versions and licenses are also available for most titles. For more information, reference our Print and eBook Bulk Sales web page at http://www.apress.com/bulk-sales.

Any source code or other supplementary material referenced by the author in this book is available to readers on GitHub via the book's product page, located at www.apress.com/978-1-4842-3563-8. For more detailed information, please visit http://www.apress.com/source-code.

Printed on acid-free paper

I would like to dedicate this book to my dear parents, Hema and Subrahmanyeswara Rao, to my lovely wife, Sindhura, and my dearest daughter, Hemanvi. This work would not have been possible without their support and encouragement.

Table of Contents

About the Author

V Kishore Ayyadevara is passionate about all things data. He has been working at the intersection of technology, data, and machine learning to identify, communicate, and solve business problems for more than a decade.

He's worked for American Express in risk management, in Amazon's supply chain analytics teams, and is currently leading data product development for a startup. In this role, he is responsible for implementing a variety of analytical solutions and building strong data science teams. He received his MBA from IIM Calcutta.

Kishore is an active learner, and his interests include identifying business problems that can be solved using data, simplifying the complexity within data science, and in transferring techniques across domains to achieve quantifiable business results.

He can be reached at www.linkedin.com/in/kishore-ayyadevara/

About the Technical Reviewer

Manohar Swamynathan is a data science practitioner and an avid programmer, with more than 13 years of experience in various data science–related areas, including data warehousing, business intelligence (BI), analytical tool development, ad-hoc analysis, predictive modeling, data science product development, consulting, formulating strategy, and executing analytics programs. He's made a career covering the lifecycle of data across different domains, including the US mortgage banking, retail/e-commerce, insurance, and industrial IoT. He has a bachelor's degree with a specialization in physics, mathematics, and computers, and a master's degree in project management. He currently lives in Bengaluru, the Silicon Valley of India.

He is the author of the book *Mastering Machine Learning with Python in Six Steps* (Apress, 2017). You can learn more about his various other activities on his website: www.mswamynathan.com.

Acknowledgments

I am grateful to my wife, Sindhura, for her love and constant support and for being a source of inspiration all through.

Sincere thanks to the Apress team, Celestin, Divya, and Matt, for their support and belief in me. Special thanks to Manohar for his review and helpful feedback. This book would not have been in this shape, without the great support from Arockia Rajan and Corbin Collins.

Thanks to Santanu Pattanayak and Antonio Gulli, who reviewed a few chapters, and also a few individuals in my organization who helped me considerably in proofreading and initial reviews: Praveen Balireddy, Arunjith, Navatha Komatireddy, Aravind Atreya, and Anugna Reddy.

Introduction

Machine learning techniques are being adopted for a variety of applications. With an increase in the adoption of machine learning techniques, it is very important for the developers of machine learning applications to understand *what* the underlying algorithms are learning, and more importantly, to understand *how* the various algorithms are learning the patterns from raw data so that they can be leveraged even more effectively.

This book is intended for data scientists and analysts who are interested in looking under the hood of various machine learning algorithms. This book will give you the confidence and skills when developing the major machine learning models and when evaluating a model that is presented to you.

True to the spirit of understanding what the machine learning algorithms are learning and how they are learning them, we first build the algorithms in Excel so that we can peek inside the black box of how the algorithms are working. In this way, the reader learns how the various levers in an algorithm impact the final result.

Once we've seen how the algorithms work, we implement them in both Python and R. However, this is *not* a book on Python or R, and I expect the reader to have some familiarity with programming. That said, the basics of Excel, Python, and R are explained in the appendix.

Chapter 1 introduces the basic terminology of data science and discusses the typical workflow of a data science project.

Chapters 2–10 cover some of the major supervised machine learning and deep learning algorithms used in industry.

Chapters 11 and 12 discuss the major unsupervised learning algorithms.

In Chapter 13, we implement the various techniques used in recommender systems to predict the likelihood of a user liking an item.

Finally, Chapter 14 looks at using the three major cloud service providers: Google Cloud Platform, Microsoft Azure, and Amazon Web Services.

All the datasets used in the book and the code snippets are available on GitHub at `https://github.com/kishore-ayyadevara/Pro-Machine-Learning`.

CHAPTER 1

Basics of Machine Learning

Machine learning can be broadly classified into supervised and unsupervised learning. By definition, the term *supervised* means that the "machine" (the system) learns with the help of something—typically a labeled training data.

Training data (or a *dataset*) is the basis on which the system learns to infer. An example of this process is to show the system a set of images of cats and dogs with the corresponding labels of the images (the labels say whether the image is of a cat or a dog) and let the system decipher the features of cats and dogs.

Similarly, *unsupervised* learning is the process of grouping data into similar categories. An example of this is to input into the system a set of images of dogs and cats without mentioning which image belongs to which category and let the system group the two types of images into different buckets based on the similarity of images.

In this chapter, we will go through the following:

- The difference between regression and classification
- The need for training, validation, and testing data
- The different measures of accuracy

Regression and Classification

Let's assume that we are forecasting for the number of units of Coke that would be sold in summer in a certain region. The value ranges between certain values—let's say 1 million to 1.2 million units per week. Typically, *regression* is a way of forecasting for such continuous variables.

© V Kishore Ayyadevara 2018
V. K. Ayyadevara, *Pro Machine Learning Algorithms*, https://doi.org/10.1007/978-1-4842-3564-5_1

Classification or *prediction*, on the other hand, predicts for events that have few distinct outcomes—for example, whether a day will be sunny or rainy.

Linear regression is a typical example of a technique to forecast continuous variables, whereas logistic regression is a typical technique to predict discrete variables. There are a host of other techniques, including decision trees, random forests, GBM, neural networks, and more, that can help predict both continuous and discrete outcomes.

Training and Testing Data

Typically, in regression, we deal with the problem of generalization/overfitting. *Overfitting* problems arise when the model is so complex that it perfectly fits *all* the data points, resulting in a minimal possible error rate. A typical example of an overfitted dataset looks like Figure 1-1.

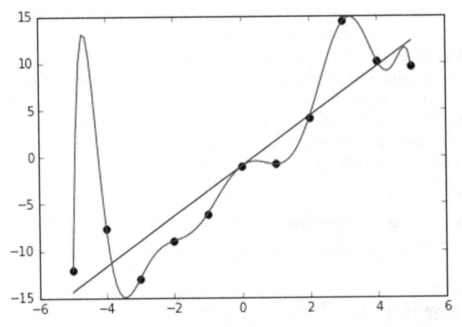

Figure 1-1. *An overfitted dataset*

From the dataset in the figure, you can see that the straight line does not fit all the data points perfectly, whereas the curved line fits the points perfectly—hence the curve has minimal error on the data points on which it is trained.

However, the straight line has a better chance of being more generalizable when compared to the curve on a new dataset. So, in practice, regression/classification is a trade-off between the generalizability of the model and complexity of model.

The lower the generalizability of the model, the higher the error rate will be on "unseen" data points.

This phenomenon can be observed in Figure 1-2. As the complexity of the model increases, the error rate of unseen data points keeps reducing up to a point, after which it starts increasing again. However, the error rate on training dataset keeps on decreasing as the complexity of model increases - eventually leading to overfitting.

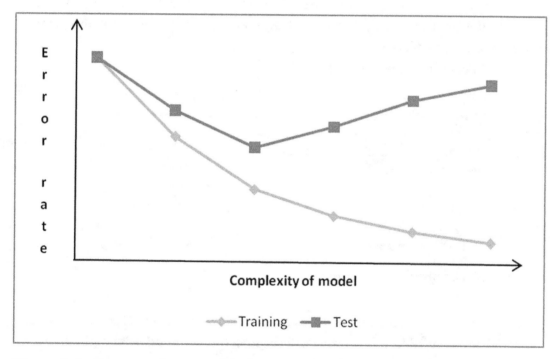

Figure 1-2. *Error rate in unseen data points*

The *unseen* data points are the points that are not used in training the model, but are used in testing the accuracy of the model, and so are called *testing data* or *test data*.

The Need for Validation Dataset

The major problem in having a fixed training and testing dataset is that the test dataset might be very similar to the training dataset, whereas a new (future) dataset might not be very similar to the training dataset. The result of a future dataset not being similar to a training dataset is that the model's accuracy for the future dataset may be very low.

3

An intuition of the problem is typically seen in data science competitions and hackathons like Kaggle (`www.kaggle.com`). The public leaderboard is not always the same as the private leaderboard. Typically, for a test dataset, the competition organizer will not tell the users which rows of the test dataset belong to the public leaderboard and which belong to the private leaderboard. Essentially, a randomly selected subset of test dataset goes to the public leaderboard and the rest goes to the private leaderboard.

One can think of the private leaderboard as a test dataset for which the accuracy is not known to the user, whereas with the public leaderboard the user is told the accuracy of the model.

Potentially, people overfit on the basis of the public leaderboard, and the private leaderboard might be a slightly different dataset that is not highly representative of the public leaderboard's dataset.

The problem can be seen in Figure 1-3.

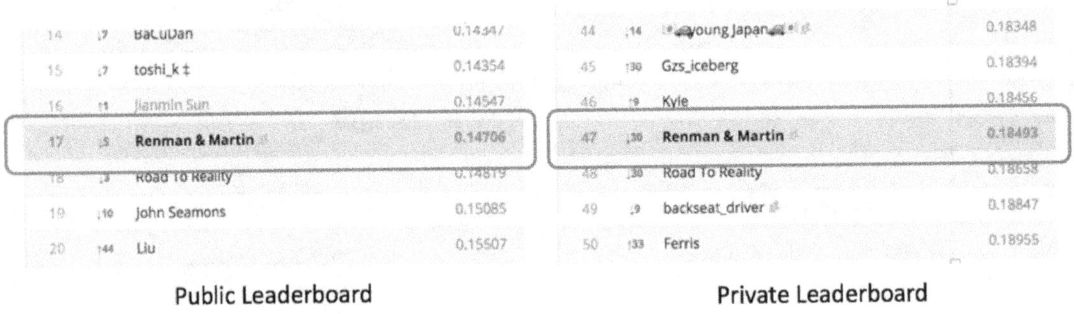

14	;7	BaCuDan	0.14347		44	;14	young Japan	0.18348
15	;7	toshi_k ‡	0.14354		45	;30	Gzs_iceberg	0.18394
16	11	Jianmin Sun	0.14547		46	;9	Kyle	0.18456
17	;5	**Renman & Martin**	0.14706		47	;30	**Renman & Martin**	0.18493
18	;3	Road To Reality	0.14819		48	;30	Road To Reality	0.18658
19	;10	John Seamons	0.15085		49	;9	backseat_driver	0.18847
20	;44	Liu	0.15507		50	;33	Ferris	0.18955

Public Leaderboard   Private Leaderboard

Figure 1-3. *The problem illustrated*

In this case, you would notice that a user moved down from rank 17 to rank 47 when compared between public and private leaderboards. *Cross-validation* is a technique that helps avoid the problem. Let's go through the workings in detail.

If we only have a training and testing dataset, given that the testing dataset would be unseen by the model, we would not be in a position to come up with the combination of *hyper-parameters* (A *hyper-parameter* can be thought of as a knob that we change to improve our model's accuracy) that maximize the model's accuracy on unseen data unless we have a third dataset. *Validation* is the third dataset that can be used to see how accurate the model is when the hyper-parameters are changed. Typically, out of the 100% data points in a dataset, 60% are used for training, 20% are used for validation, and the remaining 20% are for testing the dataset.

Another idea for a validation dataset goes like this: assume that you are building a model to predict whether a customer is likely to churn in the next two months. Most of the dataset will be used to train the model, and the rest can be used to test the dataset. But in most of the techniques we will deal with in subsequent chapters, you'll notice that they involve hyper-parameters.

As we keep changing the hyper-parameters, the accuracy of a model varies by quite a bit, but unless there is another dataset, we cannot ascertain whether accuracy is improving. Here's why:

1. We cannot test a model's accuracy on the dataset on which it is trained.

2. We cannot use the result of test dataset accuracy to finalize the ideal hyper-parameters, because, practically, the test dataset is unseen by the model.

Hence, the need for a third dataset—the validation dataset.

Measures of Accuracy

In a typical linear regression (where continuous values are predicted), there are a couple of ways of measuring the error of a model. Typically, error is measured on the testing dataset, because measuring error on the training dataset (the dataset a model is built on) is misleading—as the model has already seen the data points, and we would not be in a position to say anything about the accuracy on a future dataset if we test the model's accuracy on the training dataset only. That's why error is always measured on the dataset that is *not* used to build a model.

Absolute Error

Absolute error is defined as the absolute value of the difference between forecasted value and actual value. Let's imagine a scenario as follows:

	Actual value	Predicted value	Error	Absolute error
Data point 1	100	120	20	20
Data point 2	100	80	−20	20
Overall	200	200	0	40

In this scenario, we might incorrectly see that the overall error is 0 (because one error is +20 and the other is –20). If we assume that the overall error of the model is 0, we are missing the fact that the model is not working well on individual data points.

To avoid the issue of a positive error and negative error cancelling out each other and thus resulting in minimal error, we consider the *absolute error* of a model, which in this case is 40, and the absolute error rate is 40 / 200 = 20%

Root Mean Square Error

Another approach to solving the problem of inconsistent signs of error is to *square* the error (the square of a negative number is a positive number). The scenario under discussion above can be translated as follows:

	Actual value	Predicted value	Error	Squared error
Data point 1	100	120	20	400
Data point 2	100	80	–20	400
Overall	200	200	0	800

Now the overall squared error is 800, and the *root mean squared error* (RMSE) is the square root of (800 / 2), which is 20.

Confusion Matrix

Absolute error and RMSE are applicable while predicting continuous variables. However, predicting an event with discrete outcomes is a different process. Discrete event prediction happens in terms of *probability*—the result of the model is a probability that a certain event happens. In such cases, even though absolute error and RMSE can theoretically be used, there are other relevant metrics.

A *confusion matrix* counts the number of instances when the model predicted the outcome of an event and measures it against the actual values, as follows:

	Predicted fraud	Predicted non-fraud
Actual fraud	True positive (TP)	False negative (FN)
Actual non-fraud	False positive (FP)	True negative (TN)

- Sensitivity or true positive rate or recall = true positive / (total positives) = TP/ (TP + FN)

- Specificity or true negative rate = true negative / (total negative) = TN / (FP + TN)

- Precision or positive predicted value = TP / (TP + FP)

- Recall = TP / (TP+FN)

- Accuracy = (TP + TN) / (TP + FN + FP + TN)

- F1 score = 2TP/ (2TP + FP + FN)

AUC Value and ROC Curve

Let's say you are consulting for an operations team that manually reviews e-commerce transactions to see if they are fraud or not.

- The cost associated with such a process is the manpower required to review all the transactions.

- The benefit associated with the cost is the number of fraudulent transactions that are preempted because of the manual review.

- The overall profit associated with this setup above is the money saved by preventing fraud minus the cost of manual review.

In such a scenario, a model can come in handy as follows: we could come up with a model that gives a score to each transaction. Each transaction is scored on the probability of being a fraud. This way, all the transactions that have very little chances of being a fraud need not be reviewed by a manual reviewer. The benefit of the model thus would be to reduce the number of transactions that need to be reviewed, thereby reducing the amount of human resources needed to review the transactions and reducing the cost associated with the reviews. However, because some transactions are not reviewed, however small the probability of fraud is, there could still be some fraud that is not captured because some transactions are not reviewed.

In that scenario, a model could be helpful if it improves the overall profit by reducing the number of transactions to be reviewed (which, hopefully, are the transactions that are less likely to be fraud transactions).

The steps we would follow in calculating the area under the curve (AUC) are as follows:

1. Score each transaction to calculate the probability of fraud. (The scoring is based on a predictive model—more details on this in Chapter 3.)

2. Order the transactions in descending order of probability.

There should be very few data points that are *non-frauds* at the top of the ordered dataset and very few data points that are *frauds* at the bottom of the ordered dataset. AUC value penalizes for having such anomalies in the dataset.

For now, let's assume a total of 1,000,000 transactions are to be reviewed, and based on history, on average 1% of the total transactions are fraudulent.

- The x-axis of the receiver operating characteristic (ROC) curve is the cumulative number of points (transactions) considered.

- The y-axis is the cumulative number of fraudulent transactions captured.

Once we order the dataset, intuitively all the high-probability transactions are fraudulent transactions, and low-probability transactions are not fraudulent transactions. The cumulative number of frauds captured increases as we look at the initial few transactions, and after a certain point, it saturates as a further increase in transactions would not increase fraudulent transactions.

The graph of cumulative transactions reviewed on the x-axis and cumulative frauds captured on the y-axis would look like Figure 1-4.

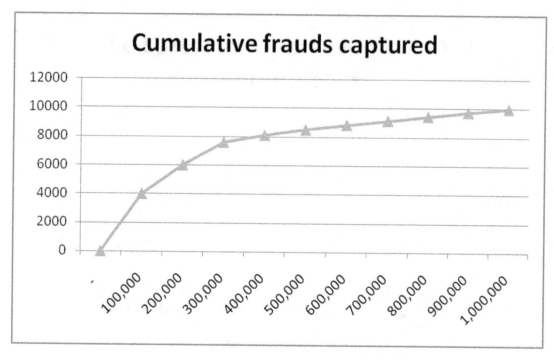

Figure 1-4. *Cumulative frauds captured when using a model*

In this scenario, we have a total of 10,000 fraudulent transactions out of a total 1,000,000 transactions. That's an average 1% fraudulent rate—that is, one out of every 100 transactions is fraudulent.

If we do not have any model, our random guess would increment slowly, as shown in Figure 1-5.

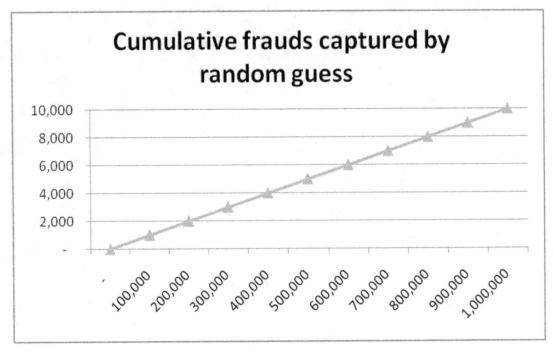

Figure 1-5. *Cumulative frauds captured when transactions are randomly sampled*

In Figure 1-5, you can see that the line divides the total dataset into two roughly equal parts—the area under the line is equal to 0.5 times of the total area. For convenience, if we assume that the total area of the plot is 1 unit, then the total area under the line generated by random guess model would be 0.5.

A comparison of the cumulative frauds captured based on the predictive model and random guess would be as shown in Figure 1-6.

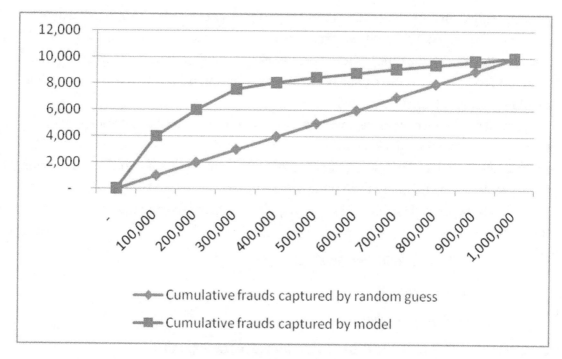

Figure 1-6. *Comparison of cumulative frauds*

Note that the area under the curve (AUC) below the curve generated by the predictive model is > 0.5 in this instance.

Thus, the higher the AUC, the better the predictive power of the model.

Unsupervised Learning

So far we have looked at supervised learning, where there is a dependent variable (the variable we are trying to predict) and an independent variable (the variable(s) we use to predict the dependent variable value).

However, in some scenarios, we would only have the independent variables—for example, in cases where we have to group customers based on certain characteristics. Unsupervised learning techniques come in handy in those cases.

There are two major types of unsupervised techniques:

- Clustering-based approach

- Principal components analysis (PCA)

Clustering is an approach where rows are grouped, and *PCA* is an approach where columns are grouped. We can think of clustering as being useful in assigning a given customer into one or the other group (because each customer typically represents a row in the dataset), whereas PCA can be useful in grouping columns (alternatively, reducing the dimensionality/variables of data).

Though clustering helps in segmenting customers, it can also be a powerful pre-processing step in our model-building process (you'll read more about that in Chapter 11). PCA can help speed up the model-building process by reducing the number of dimensions, thereby reducing the number of parameters to estimate.

In this book, we will be dealing with a majority of supervised and unsupervised algorithms as follows:

1. We first hand-code them in Excel.

2. We implement in R.

3. We implement in Python.

The basics of Excel, R and Python are outlined in the appendix.

Typical Approach Towards Building a Model

In the previous section, we saw a scenario of the cost-benefit analysis of an operations team implementing the predictive models in a real-world scenario. In this section, we'll look at some of the points you should consider while building the predictive models.

Where Is the Data Fetched From?

Typically, data is available in tables in database, CSV, or text files. In a database, different tables may be capturing different information. For example, in order to understand fraudulent transactions, we would be likely to join a transactions table with customer demographics table to derive insights from data.

Which Data Needs to Be Fetched?

The output of a prediction exercise is only as good as the inputs that go into the model. The key part in getting the input right is understanding the drivers/ characteristics of what we are trying to predict better—in our case, understanding the characteristics of a fraudulent transaction better.

Here is where a data scientist typically dons the hat of a management consultant. They research the factors that might be driving the event they are trying to predict. They could do that by reaching out to the people who are working in the front line—for example, the fraud risk investigators who are manually reviewing the transactions—to understand the key factors that they look at while investigating a transaction.

Pre-processing the Data

The input data does not always come in clean every time. There may be multiple issues that need to be handled before building a model:

- *Missing values in data*: Missing values in data exist when a variable (data point) is not recorded or when joins across different tables result in a nonexistent value.

- Missing values can be imputed in a few ways. The simplest is by replacing the missing value with the average/ median of the column. Another way to replace a missing value is to add some intelligence based on the rest of variables available in a transaction. This method is known as identifying the *K-nearest neighbors* (more on this in Chapter 13).

- *Outliers in data*: Outliers within the input variables result in inefficient optimization across the regression-based techniques (Chapter 2 talks more about the affect of outliers). Typically outliers are handled by capping variables at a certain percentile value (95%, for example).

- *Transformation of variables*: The variable transformations available are as follows:

  - *Scaling a variable*: Scaling a variable in cases of techniques based on gradient descent generally result in faster optimization.

  - *Log/Squared transformation*: Log/Squared transformation comes in handy in scenarios where the input variable shares a non-linear relation with the dependent variable.

Feature Interaction

Consider the scenario where, the chances of a person's survival *on the Titanic* is high when the person is male and also has low age. A typical regression-based technique would not take such a feature interaction into account, whereas a tree-based technique would. *Feature interaction* is the process of creating new variables based on a combination of variables. Note that, more often than not, feature interaction is known by understanding the *business* (the event that we are trying to predict) better.

Feature Generation

Feature generation is a process of finding additional features from the dataset. For example, a feature for predicting fraudulent transaction would be *time since the last transaction* for a given transaction. Such features are not available straightaway, but can only be derived by understanding the problem we are trying to solve.

Building the Models

Once the data is in place and the pre-processing steps are done, building a predictive model would be the next step. Multiple machine learning techniques would be helpful in building a predictive model. Details on the major machine learning techniques are explored in the rest of chapters.

Productionalizing the Models

Once the final model is in place, *productionalizing* a model varies, depending on the use case. For example, a data scientist can do an offline analysis looking at the historical purchases of a customer and come up with a list of products that are to be sent as recommendation over email, customized for the specific customer. In another scenario, online recommendation systems work on a real-time basis and a data scientist might have to provide the model to a data engineer who then implements the model in production to generate recommendations on a real time basis.

Build, Deploy, Test, and Iterate

In general, building a model is not a one-time exercise. You need to show the value of moving from the prior process to a new process. In such a scenario, you typically follow the A/B testing or test/control approach, where the models are deployed only for a small amount of total possible transactions/customers. The two groups are then compared to see whether the deployment of models has indeed resulted in an improvement in the metric the business is interested in achieving. Once the model shows promise, it is expanded to more total possible transactions/customers. Once consensus is reached that the model is promising, it is accepted as a final solution. Otherwise, the data scientist reiterates with the new information from the previous A/B testing experiment.

Summary

In this chapter, we looked into the basic terminology of machine learning. We also discussed the various error measures you can use in evaluating a model. And we talked about the various steps involved in leveraging machine learning algorithms to solve a business problem.

CHAPTER 2

Linear Regression

In order to understand linear regression, let's parse it:

- *Linear*: Arranged in or extending along a straight or nearly straight line, as in "linear movement."

- *Regression*: A technique for determining the statistical relationship between two or more variables where a change in one variable is caused by a change in another variable.

Combining those, we can define *linear regression* as a relationship between two variables where an increase in one variable impacts another variable to increase or decrease proportionately (that is, linearly).

In this chapter, we will learn the following:

- How linear regression works

- Common pitfalls to avoid while building linear regression

- How to build linear regression in Excel, Python, and R

Introducing Linear Regression

Linear regression helps in interpolating the value of an unknown variable (a continuous variable) based on a known value. An application of it could be, "What is the demand for a product as the price of the product is varied?" In this application, we would have to look at the demand based on historical prices and make an estimate of demand given a new price point.

Given that we are looking at history in order to estimate a new price point, it becomes a regression problem. The fact that price and demand are linearly related to each other (the higher the price, the lower the demand and vice versa) makes it a linear regression problem.

© V Kishore Ayyadevara 2018

V. K. Ayyadevara, *Pro Machine Learning Algorithms*, https://doi.org/10.1007/978-1-4842-3564-5_2

Variables: Dependent and Independent

A *dependent* variable is the value that we are predicting for, and an *independent* variable is the variable that we are using to predict a dependent variable.

For example, temperature is directly proportional to the number of ice creams purchased. As temperature increases, the number of ice creams purchased would also increase. Here temperature is the independent variable, and based on it the number of ice creams purchased (the dependent variable) is predicted.

Correlation

From the preceding example, we may notice that ice cream purchases are *directly correlated* (that is, they move in the same or opposite direction of the independent variable, temperature) with temperature. In this example, the correlation is positive: as temperature increases, ice cream sales increase. In other cases, correlation could be negative: for example, sales of an item might increase as the price of the item is decreased.

Causation

Let's flip the scenario that ice cream sales increase as temperature increases (high + ve correlation). The flip would be that temperature increases as ice cream sales increase (high + ve correlation, too).

However, intuitively we can say with confidence that temperature is *not* controlled by ice cream sales, although the reverse is true. This brings up the concept of *causation*— that is, which event influences another event. Temperature influences ice cream sales— but not vice versa.

Simple vs. Multivariate Linear Regression

We've discussed the relationship between two variables (dependent and independent). However, a dependent variable is not influenced by just one variable but by a multitude of variables. For example, ice cream sales are influenced by temperature, but they are also influenced by the price at which ice cream is being sold, along with other factors such as location, ice cream brand, and so on.

In the case of *multivariate* linear regression, some of the variables will be positively correlated with the dependent variable and some will be negatively correlated with it.

Formalizing Simple Linear Regression

Now that we have the basic terms in place, let's dive into the details of linear regression. A simple linear regression is represented as:

$$Y = a + b * X$$

- *Y* is the dependent variable that we are predicting for.

- *X* is the independent variable.

- *a* is the bias term.

- *b* is the *slope* of the variable (the weight assigned to the independent variable).

Y and X, the dependent and independent variables should be clear enough now. Let's get introduced to the bias and weight terms (*a* and *b* in the preceding equation).

The Bias Term

Let's look at the *bias* term through an example: estimating the weight of a baby by the baby's age in months. We'll assume that the weight of a baby is solely dependent on how many months old the baby is. The baby is 3 kg when born and its weight increases at a constant rate of 0.75 kg every month.

At the end of year, the chart of baby weight looks like Figure 2-1.

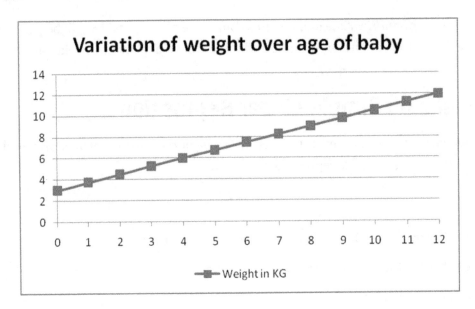

Figure 2-1. *Baby weight over time in months*

In Figure 2-1, the baby's weight starts at 3 (*a*, the bias) and increases linearly by 0.75 (*b*, the slope) every month. Note that, a bias term is the value of the dependent variable when all the independent variables are 0.

The Slope

The *slope* of a line is the difference between the x and y coordinates at both extremes of the line upon the length of line. In the preceding example, the value of slope (b) is as follows:

(Difference between y coordinates at both extremes) / (Difference between x coordinates at both extremes)

$$b = \frac{12 - 3}{(12 - 0)} = 9/12 = 0.75$$

Solving a Simple Linear Regression

We've seen a simple example of how the output of a simple linear regression might look (solving for bias and slope). In this section, we'll take the first steps towards coming up with a more generalized way to generate a regression line. The dataset provided is as follows:

Age in months	Weight in kg
0	3
1	3.75
2	4.5
3	5.25
4	6
5	6.75
6	7.5
7	8.25
8	9
9	9.75
10	10.5
11	11.25
12	12

A visualization of the data is shown in Figure 2-2.

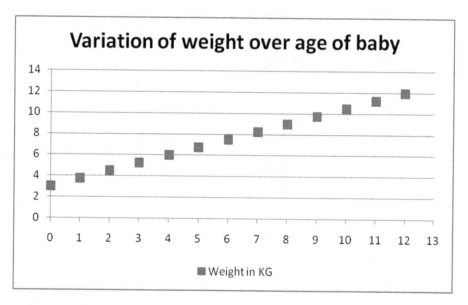

Figure 2-2. *Visualizing baby weight*

CHAPTER 2   LINEAR REGRESSION

In Figure 2-2, the x-axis is the baby's age in months, and the y-axis is the weight of the baby in a given month. For example, the weight of the baby in the first month is 3.75 kg.

Let's solve the problem from first principles. We'll assume that the dataset has only 2 data points, not 13—but, just the first 2 data points. The dataset would look like this:

Age in months	Weight in kg
0	3
1	3.75

Given that we are estimating the weight of the baby based on its age, the linear regression can be built as follows:

$$3 = a + b*(0)$$

$$3.75 = a + b*(1)$$

Solving that, we see that $a = 3$ and $b = 0.75$.

Let's apply the values of a and b on the remaining 11 data points above. The result would look like this:

Age in months	Weight In kg	Estimate of weight	Squared error of estimate
0	3	3	0
1	3.75	3.75	0
2	4.5	4.5	0
3	5.25	5.25	0
4	6	6	0
5	6.75	6.75	0
6	7.5	7.5	0
7	8.25	8.25	0
8	9	9	0
9	9.75	9.75	0
10	10.5	10.5	0
11	11.25	11.25	0
12	12	12	0
		Overall squared error	0

As you can see, the problem can be solved with minimal error rate by solving the first two data points only. However, this would likely not be the case in practice because most real data is not as clean as is presented in the table.

More General Way of Solving a Simple Linear Regression

In the preceding scenario, we saw that the coefficients are obtained by using just two data points from the total dataset—that is, we have not considered a majority of the observations in coming up with optimal *a* and *b*. To avoid leaving out most of the data points while building the equation, we can modify the objective as minimizing the overall squared error (*ordinary least squares*) across all the data points.

Minimizing the Overall Sum of Squared Error

Overall squared error is defined as the sum of the squared difference between actual and predicted values of all the observations. The reason we consider *squared* error value and not the *actual* error value is that we do not want positive error in some data points offsetting for negative error in other data points. For example, an error of +5 in three data points offsets an error of –5 in three other data points, resulting in an overall error of 0 among the six data points combined. Squared error converts the –5 error of the latter three data points into a positive number, so that the overall squared error then becomes $6 \times 5^2 = 150$.

This brings up a question: why should we minimize overall squared error? The principle is as follows:

1. Overall error is minimized if each individual data point is predicted correctly.

2. In general, overprediction by 5% is equally as bad as underprediction by 5%, hence we consider the squared error.

Let's formulate the problem:

Age in months	Weight in kg	Formula	Estimate of weight when $a = 3$ and $b = 0.75$	Squared error of estimate
0	3	$3 = a + b \times (0)$	3	0
1	3.75	$3.75 = a + b \times (1)$	3.75	0
2	4.5	$4.5 = a + b \times (2)$	4.5	0
3	5.25	$5.25 = a + b \times (3)$	5.25	0
4	6	$6 = a + b \times (4)$	6	0
5	6.75	$6.75 = a + b \times (5)$	6.75	0
6	7.5	$7.5 = a + b \times (6)$	7.5	0
7	8.25	$8.25 = a + b \times (7)$	8.25	0
8	9	$9 = a + b \times (8)$	9	0
9	9.75	$9.75 = a + b \times (9)$	9.75	0
10	10.5	$10.5 = a + b \times (10)$	10.5	0
11	11.25	$11.25 = a + b \times (11)$	11.25	0
12	12	$12 = a + b \times (12)$	12	0
			Overall squared error	0

Linear regression equation is represented in the Formula column in the preceding table.

Once the dataset (the first two columns) are converted into a *formula* (column 3), linear regression is a process of solving for the values of a and b in the formula column so that the overall squared error of estimate (the sum of squared error of all data points) is minimized.

Solving the Formula

The process of solving the formula is as simple as iterating over multiple combinations of a and b values so that the overall error is minimized as much as possible. Note that the final combination of optimal a and b value is obtained by using a technique called *gradient descent*, which is explored in Chapter 7.

Working Details of Simple Linear Regression

Solving for *a* and *b* can be understood as a *goal seek* problem in Excel, where Excel is helping identify the values of *a* and *b* that minimize the overall value.

To see how this works, look at the following dataset (available as "linear regression 101.xlsx" in github):

	A	B	D	E	F	G	H
1	Age in months	Weight in KG	Estimate of weight	Squared error of estimate			
2	0	3	2.9999951	2.40091E-11			
3	1	3.75	3.749996913	9.52768E-12		a	2.999995
4	2	4.5	4.499998727	1.62174E-12		b	0.750002
5	3	5.25	5.25000054	2.91321E-13			
6	4	6	6.000002353	5.53642E-12			
7	5	6.75	6.750004166	1.7357E-11			
8	6	7.5	7.500005979	3.57532E-11			
9	7	8.25	8.250007793	6.07248E-11			
10	8	9	9.000009606	9.2272E-11			
11	9	9.75	9.750011419	1.30395E-10			
12	10	10.5	10.50001323	1.75093E-10			
13	11	11.25	11.25001505	2.26367E-10			
14	12	12	12.00001686	2.84216E-10			
15			Overall error	1.06316E-09			

You should understand the following by checking the below in dataset:

1. How cells H3 and H4 are related to column D (estimate of weight)

2. The formula of column E

3. Cell E15, the sum of squared error for each data point

4. To obtain the optimal values of *a* and *b* (in cells H3 and H4)—go to Solver in Excel and add the following constraints:

   a. Minimize the value in cell E15

   b. By changing cells H3 and H4

	A	B	D	E	F	G	H
1	Age in months	Weight in KG	Estimate of weight	Squared error of estimate			
2	0	3	=$H$3+$H$4*A2	=(B2-D2)^2			
3	1	=B2+0.75	=$H$3+$H$4*A3	=(B3-D3)^2		a	2.99999510008889
4	2	=B3+0.75	=$H$3+$H$4*A4	=(B4-D4)^2		b	0.750001813217706
5	3	=B4+0.75	=$H$3+$H$4*A5	=(B5-D5)^2			
6	4	=B5+0.75	=$H$3+$H$4*A6	=(B6-D6)^2			
7	5	=B6+0.75	=$H$3+$H$4*A7	=(B7-D7)^2			
8	6	=B7+0.75	=$H$3+$H$4*A8	=(B8-D8)^2			
9	7	=B8+0.75	=$H$3+$H$4*A9	=(B9-D9)^2			
10	8	=B9+0.75	=$H$3+$H$4*A10	=(B10-D10)^2			
11	9	=B10+0.75	=$H$3+$H$4*A11	=(B11-D11)^2			
12	10	=B11+0.75	=$H$3+$H$4*A12	=(B12-D12)^2			
13	11	=B12+0.75	=$H$3+$H$4*A13	=(B13-D13)^2			
14	12	=B13+0.75	=$H$3+$H$4*A14	=(B14-D14)^2			
15			Overall error	=SUM(E2:E14)			

Complicating Simple Linear Regression a Little

In the preceding example, we started with a scenario where the values fit perfectly: $a = 3$ and $b = 0.75$.

The reason for zero error rate is that we defined the scenario first and then defined the approach—that is, a baby is 3 kg at birth and the weight increases by 0.75 kg every month. However, in practice the scenario is different: "Every baby is different."

Let's visualize this new scenario through a dataset (available as "Baby age to weight relation.xlsx" in github). Here, we have the age and weight measurement of two different babies.

The plot of age-to-weight relationship now looks like Figure 2-3.

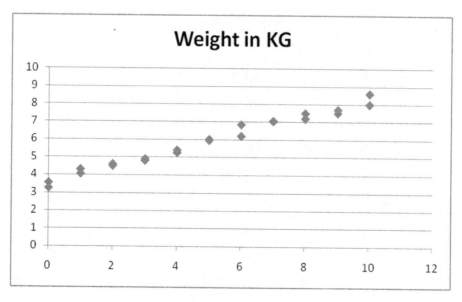

Figure 2-3. *Age-to-weight reltionship*

The value of weight increases as age increases, but not in the exact trend of starting at 3 kg and increasing by 0.75 kg every month, as seen in the simplest example.

To solve for this, we go through the same rigor we did earlier:

1. Initialize with arbitrary values of *a* and *b* (for example, each equals 1).

2. Make a new column for the forecast with the value of
    a + b × X – column C.

3. Make a new column for squared error, column D.

4. Calculate overall error in cell G7.

5. Invoke the Solver to minimize cell G7 by changing cells *a* and *b*
    —that is, G3 and G4.

	A	B	C	D	E	F	G
1	Age in months	Weight in KG	Estimate	squared error			
2	0	3.54	1	6.4516			
3	1	4.29	2	5.2441		a	1
4	2	4.59	3	2.5281		b	1
5	3	4.79	4	0.6241			
6	4	5.24	5	0.0576			
7	5	6	6	0		Overall error	61.788
8	6	6.19	7	0.6561			
9	7	7.04	8	0.9216			
10	8	7.19	9	3.2761			
11	9	7.5	10	6.25			
12	10	8.59	11	5.8081			
13	0	3.24	1	5.0176			
14	1	4.04	2	4.1616			
15	2	4.49	3	2.2201			
16	3	4.89	4	0.7921			
17	4	5.39	5	0.1521			
18	5	5.94	6	0.0036			
19	6	6.84	7	0.0256			
20	7	7.04	8	0.9216			
21	8	7.49	9	2.2801			
22	9	7.69	10	5.3361			
23	10	7.99	11	9.0601			

The cell connections in the preceding scenario are as follows:

	A	B	C	D
1	Age in months	Weight in KG	Estimate	squared error
2	0	3.54	=$G$3+$G$4*A2	=(B2-C2)^2
3	1	4.29	=$G$3+$G$4*A3	=(B3-C3)^2
4	2	4.59	=$G$3+$G$4*A4	=(B4-C4)^2
5	3	4.79	=$G$3+$G$4*A5	=(B5-C5)^2
6	4	5.24	=$G$3+$G$4*A6	=(B6-C6)^2
7	5	6	=$G$3+$G$4*A7	=(B7-C7)^2

The cell values of G3 and G4 that minimize the overall error are the optimal values of *a* and *b*.

Arriving at Optimal Coefficient Values

Optimal values of coefficients are arrived at using a technique called *gradient descent*. Chapter 7 contains a detailed discussion of how gradient descent works, but for now, let's begin to understand gradient descent using the following steps:

1. Initialize the value of coefficients (a and b) randomly.

2. Calculate the cost function—that is, the sum of squared error across all the data points in the training dataset.

3. Change the value of coefficients slightly, say, +1% of its value.

4. Check whether, by changing the value of coefficients slightly, overall squared error decreases or increases.

5. If overall squared error decreases by changing the value of coefficient by +1%, then proceed further, else reduce the coefficient by 1%.

6. Repeat steps 2–4 until overall squared error is the least.

Introducing Root Mean Squared Error

So far, we have seen that the overall error is the sum of the square of difference between forecasted and actual values for each data point. Note that, in general, as the number of data points increase, the overall squared error increases.

In order to normalize for the number of observations in data—that is, having a meaningful error measure, we would consider the square root of mean of error (as we have squared the difference while calculating error). *Root mean squared error* (RMSE) is calculated as follows (in cell G9):

	A	B	C	D	E	F	G
1	Age in months	Weight in KG	Estimate	squared error			
2	0	3.54	=$G$3+$G$4*A2	=(B2-C2)^2			
3	1	4.29	=$G$3+$G$4*A3	=(B3-C3)^2		a	1
4	2	4.59	=$G$3+$G$4*A4	=(B4-C4)^2		b	1
5	3	4.79	=$G$3+$G$4*A5	=(B5-C5)^2			
6	4	5.24	=$G$3+$G$4*A6	=(B6-C6)^2			
7	5	6	=$G$3+$G$4*A7	=(B7-C7)^2		Overall error	=SUM(D2:D23)
8	6	6.19	=$G$3+$G$4*A8	=(B8-C8)^2			
9	7	7.04	=$G$3+$G$4*A9	=(B9-C9)^2		RMSE	=SQRT(AVERAGE(D2:D23))
10	8	7.19	=$G$3+$G$4*A10	=(B10-C10)^2			
11	9	7.5	=$G$3+$G$4*A11	=(B11-C11)^2			
12	10	8.59	=$G$3+$G$4*A12	=(B12-C12)^2			
13	0	3.24	=$G$3+$G$4*A13	=(B13-C13)^2			
14	1	4.04	=$G$3+$G$4*A14	=(B14-C14)^2			
15	2	4.49	=$G$3+$G$4*A15	=(B15-C15)^2			
16	3	4.89	=$G$3+$G$4*A16	=(B16-C16)^2			
17	4	5.39	=$G$3+$G$4*A17	=(B17-C17)^2			
18	5	5.94	=$G$3+$G$4*A18	=(B18-C18)^2			
19	6	6.84	=$G$3+$G$4*A19	=(B19-C19)^2			
20	7	7.04	=$G$3+$G$4*A20	=(B20-C20)^2			
21	8	7.49	=$G$3+$G$4*A21	=(B21-C21)^2			
22	9	7.69	=$G$3+$G$4*A22	=(B22-C22)^2			
23	10	7.99	=$G$3+$G$4*A23	=(B23-C23)^2			

Note that in the preceding dataset, we would have to solve for the optimal values of a and b (cells G3 and G4) that minimize the overall error.

Running a Simple Linear Regression in R

To understand the implementation details of the material covered in the preceding sections, we'll run the linear regression in R (available as "simple linear regression.R" in github).

```
# import file
data=read.csv("D:/Pro ML book/linear_reg_example.csv")
# Build model
lm=glm(Weight~Age,data=data)
# summarize model
summary(lm)
```

The function lm stands for *linear model*, and the general syntax is as follows:

```
lm(y~x,data=data)
```

where y is the dependent variable, x is the independent variable, and data is the dataset.

summary(lm) gives a summary of the model along with the variables that came in significant, along with some automated tests. Let's parse them one at a time:

```
> summary(lm)

Call:
lm(formula = Weight ~ Age, data = data)

Residuals:
     Min       1Q    Median       3Q      Max
-0.30800 -0.16305  0.00482  0.14382  0.45618

Coefficients:
             Estimate Std. Error t value Pr(>|t|)
(Intercept)  3.53545    0.08488   41.65   <2e-16 ***
Age          0.47473    0.01435   33.09   <2e-16 ***
---
Signif. codes:  0 '***' 0.001 '**' 0.01 '*' 0.05 '.' 0.1 ' ' 1

Residual standard error: 0.2128 on 20 degrees of freedom
Multiple R-squared:  0.9821,    Adjusted R-squared:  0.9812
F-statistic:  1095 on 1 and 20 DF,  p-value: < 2.2e-16
```

Residuals

Residual is nothing but the error value (the difference between actual and forecasted value). The summary function automatically gives us the distribution of residuals. For example, consider the residuals of the model on the dataset we trained.

Distribution of residuals using the model is calculated as follows:

```
#Extracting prediction
data$prediction=predict(lm,data)
# Extracting residuals
data$residual = data$Weight - data$prediction
# summarizing the residuals
summary(data$residual)
```

In the preceding code snippet, the predict function takes the model to implement and the dataset to work on as inputs and produces the predictions as output.

Note   The output of the summary function is the various quartile values in the residual column.

Coefficients

The *coefficients* section of the output gives a summary version of the intercept and bias that got derived. (Intercept) is the bias term (*a*), and Age is the independent variable:

- Estimate is the value of *a* and *b* each.

- Std error gives us a sense of variation in the values of *a* and *b* if we draw random samples from the total population. Lower the ratio of standard error to intercept, more stable is the model.

Let's look at a way in which we can visualize/calculate the standard error values. The following steps extract the standard error value:

1. Randomly sample 50% of the total dataset.

2. Fit a lm model on the sampled data.

3. Extract the coefficient of the independent variable for the model fitted on sampled data.

4. Repeat the whole process over 100 iterations.

In code, the preceding would translate as follows:

```
# Initialize an object that stores the various coefficient values
samp_coef=c()
# Repeat the experiment 100 times
for(i in 1:100){
  # sample 50% of total data
  samp=sample(nrow(data),0.5*nrow(data))
  data2=data[samp,]
  # fit a model on the sampled data
  lm=lm(Weight~Age,data=data2)
```

```
# extract the coefficient of independent variable and store it
samp_coef=c(samp_coef,lm$coefficients['Age'])
}
sd(samp_coef)
```

Note that the lower the standard deviation, the closer the coefficient values of sample data are to the original data. This indicates that the coefficient values are stable regardless of the sample chosen.

t-value is the coefficient divided by the standard error. The higher the t-value, the better the model stability.

Consider the following example:

```
Coefficients:
                Estimate Std. Error t value Pr(>|t|)
(Intercept)     3.53545    0.08488   41.65   <2e-16 ***
Age             0.47473    0.01435   33.09   <2e-16 ***
```

The t-value corresponding to the variable Age would equal 0.47473/0.01435. (`Pr>|t|`) gives us the p-value corresponding to t-value. The lower the p-value, the better the model is. Let us look at the way in which we can derive p-value from t-value. A lookup for t-value to p-value is available in the link here: `http://www.socscistatistics.com/pvalues/tdistribution.aspx`

In our case, for the Age variable, t-value is 33.09.

Degrees of freedom = Number of rows in dataset − (Number of independent variables in model + 1) = 22 − (1 +1) = 20

Note that the +1 in the preceding formula comes from including the intercept term.

We would check for a two-tailed hypothesis and input the value of *t* and the degrees of freedom into the lookup table, and the output would be the corresponding p-value.

As a rule of thumb, if a variable has a p-value < 0.05, it is accepted as a significant variable in predicting the dependent variable. Let's look at the reason why.

If the p-value is high, it's because the corresponding t-value is low, and that's because the standard error is high when compared to the estimate, which ultimately signifies that samples drawn randomly from the population do not have similar coefficients.

In practice, we typically look at p-value as one of the guiding metrics in deciding whether to include an independent variable in a model or not.

SSE of Residuals (Residual Deviance)

The sum of squared error of residuals is calculated as follows:

```
# SSE of residuals
data$prediction = predict(lm,data)
sum((data$prediction-data$Weight)^2)
```

Residual deviance signifies the amount of deviance that one can expect after building the model. Ideally the residual deviance should be compared with null deviance—that is, how much the deviance has decreased because of building a model.

Null Deviance

A *null deviance* is the deviance expected when no independent variables are used in building the model.

The best guess of prediction, when there are no independent variables, is the average of the dependent variable itself. For example, if we say that, on average, there are $1,000 in sales per day, the best guess someone can make about a future sales value (when no other information is provided) is $1,000.

Thus, null deviance can be calculated as follows:

```
#Null deviance
data$prediction = mean(data$Weight)
sum((data$prediction-data$Weight)^2)
```

Note that, the prediction is just the mean of the dependent variable while calculating the null deviance.

R Squared

R squared is a measure of correlation between forecasted and actual values. It is calculated as follows:

1. Find the correlation between actual dependent variable and the forecasted dependent variable.

2. Square the correlation obtained in step 1—that is the R squared value.

R squared can also be calculated like this:

$$1-\left(\textit{Residual deviance / Null deviance}\right)$$

Null deviance—the deviance when we don't use any independent variable (but the bias/constant) in predicting the dependent variable—is calculated as follows:

$$\textit{null deviance} = \sum\left(Y - \hat{Y}\right)^2$$

where is the dependent variable and $\hat{Y}$ is the average of the dependent variable.

Residual deviance is the actual deviance when we use the independent variables to predict the dependent variable. It's calculated as follows:

$$\textit{residual deviance} = \sum\left(Y - \bar{y}\right)^2$$

where Y is the actual dependent variable and $\bar{y}$ is the predicted value of the dependent variable.

Essentially, R squared is high when residual deviance is much lower when compared to the null deviance.

F-statistic

F-statistic gives us a similar metric to R-squared. The way in which the F-statistic is calculated is as follows:

$$F = \left(\dfrac{\dfrac{SSE(N) - SSE(R)}{df_N - df_R}}{\dfrac{SSE(R)}{df_R}}\right)$$

where $SSE(N)$ is the null deviance, $SSE(R)$ is the residual deviance, df_N is the degrees of freedom of null deviance, and df_R is the degrees of freedom of residual deviance. The higher the F-statistic, the better the model. The higher the reduction in deviance from null deviance to residual deviance, the higher the predictability of using the independent variables in the model will be.

Running a Simple Linear Regression in Python

A linear regression can be run in Python using the following code (available as "Linear regression Python code.ipynb" in github):

```
# import relevant packages
# pandas package is used to import data
# statsmodels is used to invoke the functions that help in lm
import pandas as pd
import statsmodels.formula.api as smf

# import dataset
data = pd.read_csv('D:/Pro ML book/Linear regression/linear_reg_example.csv')

# run least squares regression
est = smf.ols(formula='Weight~Age',data=data)
est2=est.fit()
print(est2.summary())
```

The output of the preceding codes looks like this:

```
                          OLS Regression Results
===============================================================================
Dep. Variable:               Weight   R-squared:                      0.982
Model:                          OLS   Adj. R-squared:                 0.981
Method:               Least Squares   F-statistic:                    1095.
Date:              Wed, 30 May 2018   Prob (F-statistic):          6.13e-19
Time:                      20:56:31   Log-Likelihood:                3.8748
No. Observations:                22   AIC:                           -3.750
Df Residuals:                    20   BIC:                           -1.567
Df Model:                         1
Covariance Type:          nonrobust
===============================================================================
                 coef    std err          t      P>|t|      [0.025      0.975]
-------------------------------------------------------------------------------
Intercept      3.5355      0.085     41.654      0.000       3.358       3.713
Age            0.4747      0.014     33.089      0.000       0.445       0.505
```

Note that the coefficients section outputs of R and Python are very similar. However, this package has given us more metrics to study the level of prediction by default. We will look into those in more detail in a later section.

Common Pitfalls of Simple Linear Regression

The simple examples so far are to illustrate the basic workings of linear regression. Let's consider scenarios where it fails:

- *When the dependent and independent variables are not linearly related with each other throughout*: As the age of a baby increases, the weight increases, but the increase plateaus at a certain stage, after which the two values are not linearly dependent any more. Another example here would be the relation between age and height of an individual.

- *When there is an outlier among the values within independent variables*: Say there is an extreme value (a manual entry error) within a baby's age. Because our objective is to minimize the overall error while arriving at the *a* and *b* values of a simple linear regression, an extreme value in the independent variables can influence the parameters by quite a bit. You can see this work by changing the value of any age value and calculating the values of *a* and *b* that minimize the overall error. In this case, you would note that even though the overall error is low for the given values of *a* and *b*, it results in high error for a majority of other data points.

In order to avoid the first problem just mentioned, analysts typically see the relation between the two variables and determine the cut-off (segments) at which we can apply linear regression. For example, when predicting height based on age, there are distinct periods: 0–1 year, 2–4 years, 5–10, 10–15, 15–20, and 20+ years. Each stage would have a different slope for age-to-height relation. For example, growth rate in height is steep in the 0–1 phase when compared to 2–4, which is better than 5-10 phase, and so on.

To solve for the second problem mentioned, analysts typically perform one of the following tasks:

- *Normalize outliers to the 99th percentile value*: Normalizing to a 99th percentile value makes sure that abnormally high values do not influence the outcome by a lot. For example, in the example scenario from earlier, if age were mistyped as *1200* instead of *12*, it would have been *normalized* to 12 (which is among the highest values in age column).

- *Normalize but create a flag mentioning that the particular variable was normalized*: Sometimes there is good information within the extreme values. For example, while forecasting for credit limit, let us consider a scenario of nine people with an income of $500,000 and a tenth person with an income of $5,000,000 applying for a card, and a credit limit of $5,000,000 is given for each. Let us assume that the credit limit given to a person is the minimum of 10 times their income or $5,000,000. Running a linear regression on this would result in the slope being close to 10, but a number *less* than 10, because one person got a credit limit of $5,000,000 even though their income is $5,000,000. In such cases, if we have a flag that notes the $5,000,000 income person is an outlier, the slope would have been closer to 10.

Outlier flagging is a special case of multivariate regression, where there can be multiple independent variables within our dataset.

Multivariate Linear Regression

Multivariate regression, as its name suggests, involves multiple variables.

So far, in a simple linear regression, we have observed that the dependent variable is predicted based on a single independent variable. In practice, multiple variables commonly impact a dependent variable, which means multivariate is more common than a simple linear regression.

The same ice cream sales problem mentioned in the first section could be translated into a multivariate problem as follows:

Ice cream sales (the dependent variable) is dependent on the following:

- Temperature

- Weekend or not

- Price of ice cream

This problem can be translated into a mathematical model in the following way:

$$Y = a + w_1 * X_1 + w_2 * X_2$$

In that equation, w_1 is the weight associated with the first independent variable, w_2 is the weight (coefficient) associated with the second independent variable, and a is the bias term.

The values of a, w_1, and w_2 will be solved similarly to how we solved for a and b in the simple linear regression (the Solver in Excel).

The results and the interpretation of the summary of multivariate linear regression remains the same as we saw for simple linear regression in the earlier section.

A sample interpretation of the above scenario could be as follows:

$$\text{Sales of ice cream} = 2 + 0.1 \times \text{Temperature} + 0.2 \times \text{Weekend flag} - 0.5 \times \text{Price of ice cream}$$

The preceding equation is interpreted as follows: If temperature increases by 5 degrees with every other parameter remaining constant (that is, on a given day and price remains unchanged), sales of ice cream increases by $0.5.

Working details of Multivariate Linear Regression

To see how multivariate linear regression is calculated, let's go through the following example (available as "linear_multi_reg_example.xlsx" in github):

	A	B	C
1	Age	Weight	New
2	0	3.54	-0.58
3	1	4.29	0.84
4	2	4.59	-0.79
5	3	4.79	-0.92
6	4	5.24	-0.92
7	5	6	-0.87
8	6	6.19	0.04
9	7	7.04	0.76
10	8	7.19	-0.67
11	9	7.5	0.79
12	10	8.59	-0.3
13	0	3.24	-0.88
14	1	4.04	-0.67
15	2	4.49	0.18
16	3	4.89	0.01
17	4	5.39	-0.54
18	5	5.94	0.61
19	6	6.84	-0.18
20	7	7.04	0.11
21	8	7.49	0.57
22	9	7.69	-0.38
23	10	7.99	-0.73

For the preceding dataset - where Weight is the dependent variable and Age, New are independent variables, we would initialize estimate and random coefficients as follows:

	A	B	C	D	E	F	G	H
1	Age	Weight	New	estimate	squared error			
2	0	3.54	-0.58	=$H$3+$H$4*A2+$H$5*C2	=(B2-D2)^2			
3	1	4.29	0.84	=$H$3+$H$4*A3+$H$5*C3	=(B3-D3)^2		a	1
4	2	4.59	-0.79	=$H$3+$H$4*A4+$H$5*C4	=(B4-D4)^2		b	1
5	3	4.79	-0.92	=$H$3+$H$4*A5+$H$5*C5	=(B5-D5)^2		c	1
6	4	5.24	-0.92	=$H$3+$H$4*A6+$H$5*C6	=(B6-D6)^2			
7	5	6	-0.87	=$H$3+$H$4*A7+$H$5*C7	=(B7-D7)^2		overall error	=SUM(E2:E23)
8	6	6.19	0.04	=$H$3+$H$4*A8+$H$5*C8	=(B8-D8)^2			

In this case, we would iterate through multiple values of a, b, and c—that is, cells H3, H4, and H5 that minimize the values of overall squared error.

Multivariate Linear Regression in R

Multivariate linear regression can be performed in R as follows (available as "Multivariate linear regression.R" in github):

```
# import file
data=read.csv("D:/Pro ML book/Linear regression/linear_multi_reg_example.csv")
# Build model
lm=glm(Weight~Age+New,data=data)
# summarize model
summary(lm)
```

```
Call:
glm(formula = Weight ~ Age + New, data = data)

Deviance Residuals:
      Min        1Q    Median        3Q       Max
  -0.40301  -0.10267  -0.03258   0.12481   0.45824

Coefficients:
              Estimate Std. Error t value Pr(>|t|)
(Intercept)    3.58419    0.08806  40.700   <2e-16 ***
Age            0.46973    0.01426  32.937   <2e-16 ***
New            0.11553    0.07528   1.535    0.141
```

Note that we have specified multiple variables for regression by using the + symbol between the independent variables.

One interesting aspect we can note in the output would be that the New variable has a p-value that is greater than 0.05, and thus is an insignificant variable.

Typically, when a p-value is high, we test whether variable transformation or capping a variable would result in obtaining a low p-value. If none of the preceding techniques work, we might be better off excluding such variables.

Other details we can see here are calculated in a way similar to that of simple linear regression calculations in the previous sections.

Multivariate Linear Regression in Python

Similar to R, Python would also have a minor addition within the formula section to accommodate for multiple linear regression over simple linear regression:

```
# import relevant packages
# pandas package is used to import data
# statsmodels is used to inoke the functions that help in lm
import pandas as pd
import statsmodels.formula.api as smf
# import dataset
data = pd.read_csv('D:/Pro ML book/Linear regression/linear_multi_reg_
example.csv')
# run least squares regression
est = smf.ols(formula='Weight~Age+New',data=data)
est2=est.fit()
print(est2.summary())
```

Issue of Having a Non-significant Variable in the Model

A variable is *non-significant* when the p-value is high. p-value is typically high when the standard error is high compared to coefficient value. When standard error is high, it is an indication that there is a high variance within the multiple coefficients generated for multiple samples. When we have a new dataset—that is, a test dataset (which is not seen by the model while building the model)—the coefficients do not necessarily generalize for the new dataset.

This would result in a higher RMSE for the test dataset when the non-significant variable is included in the model, and typically RMSE is lower when the non-significant variable is not included in building the model.

Issue of Multicollinearity

One of the major issues to take care of while building a multivariate model is when the independent variables may be related to each other. This phenomenon is called *multicollinearity*. For example, in the ice cream example, if ice cream prices increase by 20% on weekends, the two independent variables (*price* and *weekend flag*) are correlated with each other. In such cases, one needs to be careful when interpreting the result—the assumption that the rest of the variables remain constant does not hold true anymore.

For example, we cannot assume that the only variable that changes on a weekend is the *weekend flag* anymore; we must also take into consideration that *price* also changes on a weekend. The problem translates to, at a given temperature, if the day happens to be a weekend, sales increase by 0.2 units as it is a weekend, but *decrease* by 0.1 as prices are increased by 20% during a weekend—hence, the net effect of sales is +0.1 units on a weekend.

Mathematical Intuition of Multicollinearity

To get a glimpse of the issues involved in having variables that are correlated with each other among independent variables, consider the following example (code available as "issues with correlated independent variables.R" in github):

```
# import dataset
data=read.csv("D:/Pro ML book/linear_reg_example.csv")
# Creating a correlated variable
data$correlated_age = data$Age*0.5 + rnorm(nrow(data))*0.1
cor(data$Age,data$correlated_age)
# Building a linear regression
lm=glm(Weight~Age+correlated_age,data=data)
summary(lm)
```

```
        Coefficients:
                        Estimate Std. Error t value Pr(>|t|)
        (Intercept)      3.53790    0.09005  39.287  <2e-16 ***
        Age              0.44522    0.27881   1.597   0.127
        correlated_age   0.05836    0.55073   0.106   0.917
```

Note that, even though *Age* is a significant variable in predicting *Weight* in the earlier examples, when a correlated variable is present in the dataset, *Age* turns out to be a non-significant variable, because it has a high p-value.

The reason for high variance in the coefficients of *Age* and *correlated_age* by sample of data is that, more often than not, although the *Age* and *correlated_age* variables are correlated, the combination of age and correlated age (when treated as a single variable—say, the average of the two variables) would have less variance in coefficients.

Given that we are using two variables, depending on the sample, *Age* might have high coefficient, and *correlated_age* might have a low coefficient, and vice versa for some other samples, resulting in a high variance in coefficients for both variables by the sample chosen.

Further Points to Consider in Multivariate Linear Regression

- *It is not advisable for a regression to have very high coefficients*: Although a regression can have high coefficients in some cases, in general, a high value of a coefficient results in a huge swing in the predicted value, even if the independent variable changes by 1 unit. For example, if *sales* is a function of *price*, where sales = 1,000,000 – 100,000 x price, a unit change of price can drastically reduce sales. In such cases, to avoid this problem, it is advisable to reduce the value of *sales* by changing it to *log(sales)* instead of *sales*, or normalize sales variable, or penalize the model for having high magnitude of weights through L1 and L2 regularizations (More on L1/ L2 regularizations in Chapter 7). This way, the *a* and *b* values in the equation remain small.

- *A regression should be built on considerable number of observations*: In general, the higher the number of data points, more reliable the model is. Moreover, the higher the number of independent variables, the more data points to consider. If we have only two data points and two independent variables, we can always come up with an equation that is perfect for the two data points. But the generalization of the equation built on two data points only is questionable. In practice, it is advisable to have the number of data points be at least 100 times the number of independent variables.

The problem of low number of rows, or high number of columns, or both, brings us to the problem of adjusted R squared. As detailed earlier, the more independent variables in an equation, the higher the chances are of fitting it closest to the dependent variable, and thus a high R squared, even if the independent variables are non-significant. Thus, there should be a way of penalizing for having a high number of independent variables over a fixed set of data points. Adjusted R squared considers the number of independent variables used in an equation and penalizes for having more independent variables. The formula for adjusted R squared is as follows:

$$R^2_{adj} = 1 - \left[\frac{(1-R^2)(n-1)}{n-k-1} \right]$$

where n is the number of data points in dataset and k is the number of independent variables in the dataset.

The model with the least adjusted R squared is generally the better model to go with.

Assumptions of Linear Regression

The assumptions of linear regression are as follows:

- *The independent variables must be linearly related to dependent variable*: If the level of linearity changes over segment, a linear model is built per segment.

- *There should not be any outliers in values among independent variables*: If there are any outliers, they should either be capped or a new variable needs to be created that flags the data points that are outliers.

- *Error values should be independent of each other*: In a typical ordinary least squares method, the error values are distributed on both sides of the fitted line (that is, some predictions will be above actuals and some will be below actuals), as shown in Figure 2-4. A linear regression cannot have errors that are all on the same side, or that follow a pattern where low values of independent variable have error of one sign while high values of independent variable have error of the opposite sign.

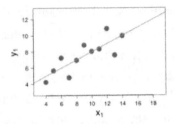

Figure 2-4. *Errors on both sides of the line*

- *Homoscedasticity*: Errors cannot get larger as the value of an independent variable increases. Error distribution should look more like a cylinder than a cone in linear regression (see Figure 2-5). In a practical scenario, we can think of the predicted value being on the x-axis and the actual value being on the y-axis.

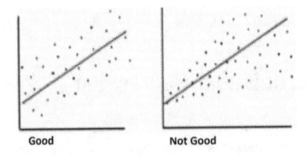

Figure 2-5. *Comparing error distributions*

- *Errors should be normally distributed*: There should be only a few data points that have high error. A majority of data points should have low error, and a few data points should have positive and negative error—that is, errors should be normally distributed (both to the left of overforecasting and to the right of underforecasting), as shown in Figure 2-6.

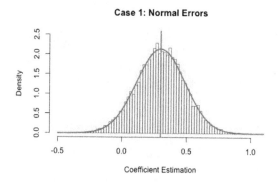

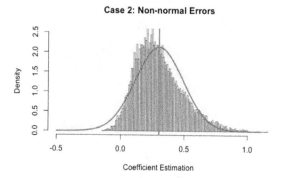

Figure 2-6. *Comparing curves*

Note   In Figure 2-6, had we adjusted the bias (intercept) in the right-hand chart slightly, more observations would now surround zero error.

Summary

In this chapter, we have learned the following:

- The sum of squared error (SSE) is the optimization based on which the coefficients in a linear regression are calculated.

- Multicollinearity is an issue when multiple independent variables are correlated to each other.

- p-value is an indicator of the significance of a variable in predicting a dependent variable.

- For a linear regression to work, the five assumptions - that is, linear relation between dependent and independent variables, no outliers, error value independence, homoscedasticity, normal distribution of errors should be satisfied.

CHAPTER 3

Logistic Regression

In Chapter 2, we looked at ways in which a variable can be estimated based on an independent variable. The dependent variables we were estimating were continuous (sales of ice cream, weight of baby). However, in most cases, we need to be forecasting or predicting for *discrete* variables—for example, whether a customer will churn or not, or whether a match will be won or not. These are the events that do not have a lot of distinct values. They have only a 1 or a 0 outcome—whether an event has happened or not.

Although a linear regression helps in forecasting the value (magnitude) of a variable, it has limitations when predicting for variables that have just two distinct classes (1 or 0). *Logistic regression* helps solve such problems, where there are a limited number of distinct values of a dependent variable.

In this chapter, we will learn the following:

- The difference between linear and logistic regression

- Building a logistic regression in Excel, R, and Python

- Ways to measure the performance of a logistic regression model

Why Does Linear Regression Fail for Discrete Outcomes?

In order to understand this, let's take a hypothetical case: predicting the result of a chess game based on the difference between the Elo ratings of the players.

Difference in rating between black and white piece players	White won?
200	0
−200	1
300	0

© V Kishore Ayyadevara 2018
V. K. Ayyadevara, *Pro Machine Learning Algorithms*, https://doi.org/10.1007/978-1-4842-3564-5_3

In the preceding simple example, if we apply linear regression, we will get the following equation:

White won = 0.55 – 0.00214 × (Difference in rating between black and white)

Let's use that formula to extrapolate on the preceding table:

Difference in rating between black and white	White won?	Prediction of linear regression
200	0	0.11
–200	1	0.97
300	0	–0.1

As you can see, the difference of 300 resulted in a prediction of less than 0. Similarly for a difference of –300, the prediction of linear regression will be beyond 1. However, in this case, values beyond 0 or 1 don't make sense, because a win is a discrete value (either 0 or 1).

Hence the predictions should be bound to either 0 to 1 only—any prediction above 1 should be capped at 1 and any prediction below 0 should be floored at 0.

This translates to a fitted line, as shown in Figure 3-1.

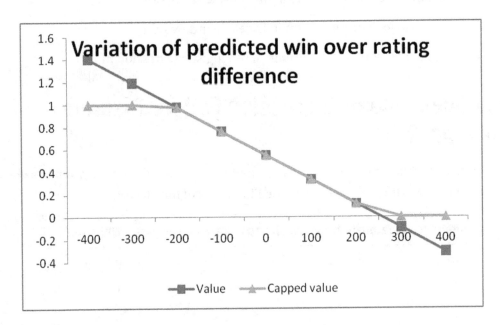

Figure 3-1. *The fitted line*

Figure 3.1 shows the following major limitations of linear regression in predicting discrete (binary in this case) variables:

- *Linear regression assumes that the variables are linearly related:* However, as player strength difference increases, chances of win vary exponentially.

- *Linear regression does not give a chance of failure:* In practice, even if there is a difference of 500 points, there is an outside chance (let's say a 1% chance) that the inferior player might win. But if capped using linear regression, there is no chance that the other player could win. In general, linear regression does not tell us the probability of an event happening after certain range.

- *Linear regression assumes that the probability increases proportionately as the independent variable increases:* The probability of win is high irrespective of whether the rating difference is +400 or +500 (as the difference is significant). Similarly, the probability of win is low, irrespective of whether the difference is –400 or –500.

A More General Solution: Sigmoid Curve

As mentioned, the major problem with linear regression is that it assumes that all relations are linear, although in practice very few are.

To solve for the limitations of linear regression, we will explore a curve called a *sigmoid curve*. The curve looks like Figure 3-2.

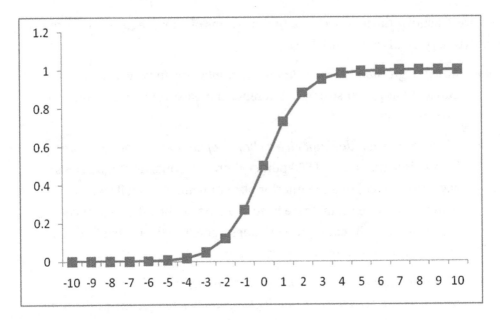

Figure 3-2. *A sigmoid curve*

The features of the sigmoid curve are as follows:

- It varies between the values 0 and 1

- It plateaus after a certain threshold (after the value 3 or -3 in Figure 3-2)

The sigmoid curve would help us solve the problems faced with linear regression—that the probability of win is high irrespective of the difference in rating between white and black piece player being +400 or +500 and that the probability of win is low, irrespective of the difference being –400 or –500.

Formalizing the Sigmoid Curve (Sigmoid Activation)

We've seen that sigmoid curve is in a better position to explain discrete phenomenon than linear regression.

A sigmoid curve can be represented in a mathematical formula as follows:

$$S(t) = \frac{1}{1 + e^{-t}}$$

In that equation, the higher the value of t, lower the value of e^{-t}, hence $S(t)$ is close to 1. And the lower the value of t (let's say –100), the higher the value of e^{-t} and the higher the value of $(1 + e^{-t})$, hence $S(t)$ is very close to 0.

From Sigmoid Curve to Logistic Regression

Linear regression assumes a linear relation between dependent and independent variables. It is written as $Y = a + b \times X$. Logistic regression moves away from the constraint that all relations are linear by applying a sigmoid curve.

Logistic regression is mathematically modeled as follows:

$$Y = \frac{1}{\left(1 + e^{-(a + b^*X)}\right)}$$

We can see that logistic regression uses independent variables in the same way as linear regression but passes them through a sigmoid activation so that the outputs are bound between 0 and 1.

In case of the presence of multiple independent variables, the equation translates to a multivariate linear regression passing through a sigmoid activation.

Interpreting the Logistic Regression

Linear regression can be interpreted in a straightforward way: as the value of the independent variable increases by 1 unit, the *output* (dependent variable) increases by b units.

To see how the output changes in a logistic regression, let's look at an example. Let's assume that the logistic regression curve we've built (we'll look at how to build a logistic regression in upcoming sections) is as follows:

$$Y = \frac{1}{\left(1 + e^{-(2 + 3^*X)}\right)}$$

- If $X = 0$, the value of $Y = 1 / (1 + \exp(-(2))) = 0.88$.

- If X is increased by 1 unit (that is, $X = 1$), the value of Y is $Y = 1 / (1 + \exp(-(2 + 3 \times 1))) = 1 / (1 + \exp(-(5))) = 0.99$.

As you see, the value of Y changed from 0.88 to 0.99 as X changed from 0 to 1. Similarly, if X were –1, Y would have been at 0.27. If X were 0, Y would have been at 0.88. There was a drastic change in Y from 0.27 to 0.88 when X went from –1 to 0 but not so drastic when X moved from 0 to 1.

Thus the impact on Y of a unit change in X depends on the equation.

The value 0.88 when X = 0 can be interpreted as the *probability*. In other words, on average in 88% of cases, the value of Y is 1 when X = 0.

Working Details of Logistic Regression

To see how a logistic regression works, we'll go through the same exercise we did to learn linear regression in the last chapter: we'll build a logistic regression equation in Excel. For this exercise, we'll use the Iris dataset. The challenge is to be able to predict whether the species is *Setosa* or not, based on a few variables (sepal, petal length, and width).

The following dataset contains the independent and dependent variable values for the exercise we are going to perform (available as "iris sample estimation.xlsx" dataset in github):

	A	B	C	D	E
1	Slength	Swidth	Plength	Pwidth	Setosa
2	5.1	3.5	1.4	0.2	1
3	4.9	3	1.4	0.2	1
4	4.7	3.2	1.3	0.2	1
5	4.6	3.1	1.5	0.2	1
6	5	3.6	1.4	0.2	1
7	5.4	3.9	1.7	0.4	1
8	4.6	3.4	1.4	0.3	1
9	5	3.4	1.5	0.2	1
10	4.4	2.9	1.4	0.2	1
11	7	3.2	4.7	1.4	0
12	6.4	3.2	4.5	1.5	0
13	6.9	3.1	4.9	1.5	0
14	5.5	2.3	4	1.3	0
15	6.5	2.8	4.6	1.5	0
16	5.7	2.8	4.5	1.3	0
17	6.3	3.3	4.7	1.6	0
18	4.9	2.4	3.3	1	0
19	6.6	2.9	4.6	1.3	0

1. Initialize the weights of independent variables to random values (let's say 1 each).

2. Once the *weights* and the *bias* are initialized, we'll estimate the *output value* (the probability of the species being *Setosa*) by applying sigmoid activation on the multivariate linear regression of independent variables.

The next table contains information about the $(a + b \times X)$ part of the sigmoid curve and ultimately the sigmoid activation value.

	A	B	C	D	E	F	G	H	I	J
1	Slength	Swidth	Plength	Pwidth	Setosa	a+b*x part of estimation	sigmoid activation			
2	5.1	3.5	1.4	0.2	1	11.2	0.999986326			
3	4.9	3	1.4	0.2	1	10.5	0.999972464			
4	4.7	3.2	1.3	0.2	1	10.4	0.999969568		a	1
5	4.6	3.1	1.5	0.2	1	10.4	0.999969568		b	1
6	5	3.6	1.4	0.2	1	11.2	0.999986326		c	1
7	5.4	3.9	1.7	0.4	1	12.4	0.999995881		d	1
8	4.6	3.4	1.4	0.3	1	10.7	0.999977456		e	1
9	5	3.4	1.5	0.2	1	11.1	0.999984888			
10	4.4	2.9	1.4	0.2	1	9.9	0.999949828			
11	7	3.2	4.7	1.4	0	17.3	0.999999969			
12	6.4	3.2	4.5	1.5	0	16.6	0.999999938			
13	6.9	3.1	4.9	1.5	0	17.4	0.999999972			
14	5.5	2.3	4	1.3	0	14.1	0.999999248			
15	6.5	2.8	4.6	1.5	0	16.4	0.999999925			
16	5.7	2.8	4.5	1.3	0	15.3	0.999999773			
17	6.3	3.3	4.7	1.6	0	16.9	0.999999954			
18	4.9	2.4	3.3	1	0	12.6	0.999996628			
19	6.6	2.9	4.6	1.3	0	16.4	0.999999925			
20										

The formula for how the values in the preceding table are obtained is given in the following table:

	A	B	C	D	E	F	G	H	I	J
1	Slength	Swidth	Plength	Pwidth	Setosa	a+b*x part of estimation	sigmoid activation			
2	5.1	3.5	1.4	0.2	1	=$J$4+$J$5*A2+$J$6*B2+$J$7*C2+$J$8*D2	=IF(F2>500,500,IF(F2<-500,-500,1/(1+EXP(-F2))))			
3	4.9	3	1.4	0.2	1	=$J$4+$J$5*A3+$J$6*B3+$J$7*C3+$J$8*D3	=IF(F3>500,500,IF(F3<-500,-500,1/(1+EXP(-F3))))			
4	4.7	3.2	1.3	0.2	1	=$J$4+$J$5*A4+$J$6*B4+$J$7*C4+$J$8*D4	=IF(F4>500,500,IF(F4<-500,-500,1/(1+EXP(-F4))))	a	1	
5	4.6	3.1	1.5	0.2	1	=$J$4+$J$5*A5+$J$6*B5+$J$7*C5+$J$8*D5	=IF(F5>500,500,IF(F5<-500,-500,1/(1+EXP(-F5))))	b	1	
6	5	3.6	1.4	0.2	1	=$J$4+$J$5*A6+$J$6*B6+$J$7*C6+$J$8*D6	=IF(F6>500,500,IF(F6<-500,-500,1/(1+EXP(-F6))))	c	1	
7	5.4	3.9	1.7	0.4	1	=$J$4+$J$5*A7+$J$6*B7+$J$7*C7+$J$8*D7	=IF(F7>500,500,IF(F7<-500,-500,1/(1+EXP(-F7))))	d	1	
8	4.6	3.4	1.4	0.3	1	=$J$4+$J$5*A8+$J$6*B8+$J$7*C8+$J$8*D8	=IF(F8>500,500,IF(F8<-500,-500,1/(1+EXP(-F8))))	e	1	
9	5	3.4	1.5	0.2	1	=$J$4+$J$5*A9+$J$6*B9+$J$7*C9+$J$8*D9	=IF(F9>500,500,IF(F9<-500,-500,1/(1+EXP(-F9))))			
10	4.4	2.9	1.4	0.2	1	=$J$4+$J$5*A10+$J$6*B10+$J$7*C10+$J$8*D10	=IF(F10>500,500,IF(F10<-500,-500,1/(1+EXP(-F10))))			
11	7	3.2	4.7	1.4	0	=$J$4+$J$5*A11+$J$6*B11+$J$7*C11+$J$8*D11	=IF(F11>500,500,IF(F11<-500,-500,1/(1+EXP(-F11))))			
12	6.4	3.2	4.5	1.5	0	=$J$4+$J$5*A12+$J$6*B12+$J$7*C12+$J$8*D12	=IF(F12>500,500,IF(F12<-500,-500,1/(1+EXP(-F12))))			
13	6.9	3.1	4.9	1.5	0	=$J$4+$J$5*A13+$J$6*B13+$J$7*C13+$J$8*D13	=IF(F13>500,500,IF(F13<-500,-500,1/(1+EXP(-F13))))			
14	5.5	2.3	4	1.3	0	=$J$4+$J$5*A14+$J$6*B14+$J$7*C14+$J$8*D14	=IF(F14>500,500,IF(F14<-500,-500,1/(1+EXP(-F14))))			
15	6.5	2.8	4.6	1.5	0	=$J$4+$J$5*A15+$J$6*B15+$J$7*C15+$J$8*D15	=IF(F15>500,500,IF(F15<-500,-500,1/(1+EXP(-F15))))			
16	5.7	2.8	4.5	1.3	0	=$J$4+$J$5*A16+$J$6*B16+$J$7*C16+$J$8*D16	=IF(F16>500,500,IF(F16<-500,-500,1/(1+EXP(-F16))))			
17	6.3	3.3	4.7	1.6	0	=$J$4+$J$5*A17+$J$6*B17+$J$7*C17+$J$8*D17	=IF(F17>500,500,IF(F17<-500,-500,1/(1+EXP(-F17))))			
18	4.9	2.4	3.3	1	0	=$J$4+$J$5*A18+$J$6*B18+$J$7*C18+$J$8*D18	=IF(F18>500,500,IF(F18<-500,-500,1/(1+EXP(-F18))))			
19	6.6	2.9	4.6	1.3	0	=$J$4+$J$5*A19+$J$6*B19+$J$7*C19+$J$8*D19	=IF(F19>500,500,IF(F19<-500,-500,1/(1+EXP(-F19))))			

The `ifelse` condition in the preceding sigmoid activation column is used only because Excel has limitations in calculating any value greater than $\exp(500)$—hence the clipping.

Estimating Error

In Chapter 2, we considered *least squares* (the squared difference) between actual and forecasted value to estimate overall error. In logistic regression, we will use a different error metric, called cross entropy.

Cross entropy is a measure of difference between two different distributions - actual distribution and predicted distribution. In order to understand cross entropy, let's see an example: two parties contest in an election, where party A won. In one scenario, chances of winning are 0.5 for each party—in other words, few conclusions can be drawn, and the information is minimal. But if party A has an 80% chance of winning, and party B has a 20% chance of winning, we can draw a conclusion about the outcome of election, as the distributions of actual and predicted values are closer.

The formula for cross entropy is as follows:

$$-\left(y\log_2 p + (1-y)\log_2 (1-p)\right)$$

where y is the actual outcome of the event and p is the predicted outcome of the event. Let's plug the two election scenarios into that equation.

Scenario 1

In this scenario, the model predicted 0.5 probability of win for party A, and the actual result of party A is 1:

Model prediction for party A	Actual outcome for party A
0.5	1

The cross entropy of this model is the following:

$$-\left(1log_2 0.5+(1-1)log_2\left(1-0.5\right)\right)=1$$

Scenario 2

In this scenario, the model predicted 0.8 probability of win for party A, and the actual result of party A is 1:

Model prediction for party A	Actual outcome for party A
0.8	1

The cross entropy of this model is the following:

$$-\left(1log_2 0.8+(1-1)log_2\left(1-0.8\right)\right)=0.32$$

We can see that scenario 2 has lower cross entropy when compared to scenario 1.

Least Squares Method and Assumption of Linearity

Given that in preceding example, when probability was 0.8, cross entropy was lower compared to when probability was 0.5, could we not have used least squares difference between predicted probability, actual value and proceeded in a similar way to how we proceeded for linear regression? This section discusses choosing cross entropy error over the least squares method.

A typical example of logistic regression is its application in predicting whether a cancer is benign or malignant based on certain attributes.

Let's compare the two cost functions (least squares method and entropy cost) in cases where the dependent variable (malignant cancer) is 1:

	A	B	C	D
1	Actual	Predicted	Squared error	Cross entropy
2	1	0.01	0.98	6.64
3	1	0.1	0.81	3.32
4	1	0.2	0.64	2.32
5	1	0.3	0.49	1.74
6	1	0.4	0.36	1.32
7	1	0.5	0.25	1.00
8	1	0.6	0.16	0.74
9	1	0.7	0.09	0.51
10	1	0.8	0.04	0.32
11	1	0.9	0.01	0.15
12	1	0.99	0.00	0.01

The formulas to obtain the preceding table are as follows:

	A	B	C	D
1	Actual	Predicted	Squared error	Cross entropy
2	1	0.01	=(A2-B2)^2	=-(A2*LOG(B2,2)+(1-A2)*LOG(1-B2,2))
3	1	0.1	=(A3-B3)^2	=-(A3*LOG(B3,2)+(1-A3)*LOG(1-B3,2))
4	1	0.2	=(A4-B4)^2	=-(A4*LOG(B4,2)+(1-A4)*LOG(1-B4,2))
5	1	0.3	=(A5-B5)^2	=-(A5*LOG(B5,2)+(1-A5)*LOG(1-B5,2))
6	1	0.4	=(A6-B6)^2	=-(A6*LOG(B6,2)+(1-A6)*LOG(1-B6,2))
7	1	0.5	=(A7-B7)^2	=-(A7*LOG(B7,2)+(1-A7)*LOG(1-B7,2))
8	1	0.6	=(A8-B8)^2	=-(A8*LOG(B8,2)+(1-A8)*LOG(1-B8,2))
9	1	0.7	=(A9-B9)^2	=-(A9*LOG(B9,2)+(1-A9)*LOG(1-B9,2))
10	1	0.8	=(A10-B10)^2	=-(A10*LOG(B10,2)+(1-A10)*LOG(1-B10,2))
11	1	0.9	=(A11-B11)^2	=-(A11*LOG(B11,2)+(1-A11)*LOG(1-B11,2))
12	1	0.99	=(A12-B12)^2	=-(A12*LOG(B12,2)+(1-A12)*LOG(1-B12,2))

Note that cross entropy penalizes heavily for high prediction errors compared to squared error: lower error values have similar loss in both squared error and cross entropy error, but for higher differences between actual and predicted values, cross entropy penalizes more than the squared error method. Thus, we will stick to cross entropy error as our error metric, preferring it to squared error for discrete variable prediction.

For the *Setosa* classification problem mentioned earlier, let's use cross entropy error instead of squared error, as follows:

	A	B	C	D	E	F	G	H	I	L	M	N
1	Slength	Swidth	Plength	Pwidth	Setosa	a+b*x part of estimation	sigmoid activation	Cross entropy error				
2	5.1	3.5	1.4	0.2	1	11.2	1.00	0.00				
3	4.9	3	1.4	0.2	1	10.5	1.00	0.00				
4	4.7	3.2	1.3	0.2	1	10.4	1.00	0.00			a	1
5	4.6	3.1	1.5	0.2	1	10.4	1.00	0.00			b	1
6	5	3.6	1.4	0.2	1	11.2	1.00	0.00			c	1
7	5.4	3.9	1.7	0.4	1	12.4	1.00	0.00			d	1
8	4.6	3.4	1.4	0.3	1	10.7	1.00	0.00			e	1
9	5	3.4	1.5	0.2	1	11.1	1.00	0.00				
10	4.4	2.9	1.4	0.2	1	9.9	1.00	0.00				
11	7	3.2	4.7	1.4	0	17.3	1.00	24.96				
12	6.4	3.2	4.5	1.5	0	16.6	1.00	23.95				
13	6.9	3.1	4.9	1.5	0	17.4	1.00	25.10				
14	5.5	2.3	4	1.3	0	14.1	1.00	20.34				
15	6.5	2.8	4.6	1.5	0	16.4	1.00	23.66				
16	5.7	2.8	4.5	1.3	0	15.3	1.00	22.07				
17	6.3	3.3	4.7	1.6	0	16.9	1.00	24.38				
18	4.9	2.4	3.3	1	0	12.6	1.00	18.18				
19	6.6	2.9	4.6	1.3	0	16.4	1.00	23.66				
20							Overall error	206.31				

Now that we have set up our problem, let's vary the parameters in such a way that overall error is minimized. This step again is performed by *gradient descent*, which can be done by using the Solver functionality in Excel.

Running a Logistic Regression in R

Now that we have some background in logistic regression, we'll dive into the implementation details of the same in R (available as "logistic regression.R" in github):

```
# import dataset
data=read.csv("D:/Pro ML book/Logistic regression/iris_sample.csv")
# build a logistic regression model
lm=glm(Setosa~.,data=data,family=binomial(logit))
# summarize the model
summary(lm)
```

The second line in the preceding code specifies that we will be using the glm (generalized linear models), in which binomial family is considered. Note that by specifying "~." we're making sure that all variables are being considered as independent variables.

summary of the logistic model gives a high-level summary similar to the way we got summary results in linear regression:

```
Call:
glm(formula = Setosa ~ ., family = binomial(logit), data = data)

Deviance Residuals:
       Min          1Q       Median          3Q         Max
-1.197e-05   -2.110e-08    0.000e+00   9.930e-07    1.292e-05

Coefficients:
              Estimate Std. Error z value Pr(>|z|)
(Intercept) -2.425e+01  4.704e+05       0        1
Slength      1.755e+00  4.541e+05       0        1
Swidth       1.570e+01  5.191e+05       0        1
Plength      4.451e+00  7.499e+05       0        1
Pwidth      -6.007e+01  1.384e+06       0        1

(Dispersion parameter for binomial family taken to be 1)

    Null deviance: 2.4953e+01  on 17  degrees of freedom
Residual deviance: 4.1464e-10  on 13  degrees of freedom
AIC: 10
```

Running a Logistic Regression in Python

Now let's see how a logistic regression equation is built in Python (available as "logistic regression.ipynb" in github):

```
# import relevant packages
# pandas package is used to import data
# statsmodels is used to invoke the functions that help in lm
import pandas as pd
import statsmodels.formula.api as smf
```

Once we import the package, we use the `logit` method in case of logistic regression, as follows:

```
# import dataset
data = pd.read_csv('D:/Pro ML book/Logistic regression/iris_sample.csv')
# run regression
est = smf.logit(formula='Setosa~Slength+Swidth+Plength+Pwidth',data=data)
est2=est.fit()
print(est2.summary())
```

The `summary` function in the preceding code gives a summary of the model, similar to the way in which we obtained summary results in linear regression.

Identifying the Measure of Interest

In linear regression, we have looked at root mean squared error (RMSE) as a way to measure error.

In logistic regression, the way we measure the performance of the model is different from how we measured it in linear regression. Let's explore why linear regression error metrics cannot be used in logistic regression.

We'll look at building a model to predict a fraudulent transaction. Let's say 1% of the total transactions are fraudulent transactions. We want to predict whether a transaction is likely to be fraud. In this particular case, we use logistic regression to predict the dependent variable *fraud transaction* by using a set of independent variables.

Why can't we use an accuracy measure? Given that only 1% of all the transactions are fraud, let's consider a scenario where all our predictions are 0. In this scenario, our model has an accuracy of 99%. But the model is not at all useful in reducing fraudulent transactions because it predicts that every transaction is not a fraud.

In a typical real-world scenario, we would build a model that predicts whether the transaction is likely to be a fraud or not, and only the transactions that have a high likelihood of fraud are flagged. The transactions that are flagged are then sent for manual review to the operations team, resulting in a lower fraudulent transaction rate.

Although we are reducing the fraud transaction rate by getting the high-likelihood transactions reviewed by the operations team, we are incurring an additional cost of manpower, because humans are required to review the transaction.

A fraud transaction prediction model can help us narrow the number of transactions that need to be reviewed by a human (operations team). Let's say in total there are a total of 1,000,000 transactions. Of those million transactions, 1% are fraudulent—so, a total of 10,000 transactions are fraudulent.

In this particular case, if there were no model, on average 1 in 100 transactions is fraudulent. The performance of the random guess model is shown in the following table:

No. of transactions reviewed	Cumulative frauds captured by random guess
-	-
100,000	1,000
200,000	2,000
300,000	3,000
400,000	4,000
500,000	5,000
600,000	6,000
700,000	7,000
800,000	8,000
900,000	9,000
1,000,000	10,000

If we were to plot that data, it would look something like Figure 3-3.

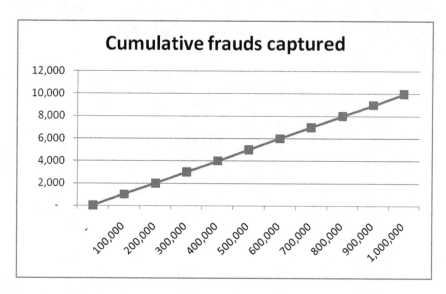

Figure 3-3. *Cumulative frauds captured by random guess model*

Now let's look at how building a model can help. We'll create a simple example to come up with an error measure:

1. Take the dataset as input and compute the probabilities of each transaction id:

Transaction id	Actual fraud	Probability of fraud
1	1	0.56
2	0	0.7
3	1	0.39
4	1	0.55
5	1	0.03
6	0	0.84
7	0	0.05
8	0	0.46
9	0	0.86
10	1	0.11

2. Sort the dataset by probability of fraud from the highest to least. The intuition is that the model performs well when there are more 1s of "Actual fraud" at the top of dataset after sorting in descending order by probability:

Transaction id	Actual fraud	Probability of fraud
9	0	0.86
6	0	0.84
2	0	0.7
1	1	0.56
4	1	0.55
8	0	0.46
3	1	0.39
10	1	0.11
7	0	0.05
5	1	0.03

3. Calculate the cumulative number of transactions captured from the sorted table:

Transaction id	Actual fraud	Probability of fraud	Cumulative transactions reviewed	Cumulative frauds captured
9	0	0.86	1	0
6	0	0.84	2	0
2	0	0.7	3	0
1	1	0.56	4	1
4	1	0.55	5	2
8	0	0.46	6	2
3	1	0.39	7	3
10	1	0.11	8	4
7	0	0.05	9	4
5	1	0.03	10	5

In this scenario, given that 5 out of 10 transactions are fraudulent, on average 1 in 2 transactions are fraudulent. So, cumulative frauds captured by using the model versus by using a random guess would be as follows:

Transaction id	Actual fraud	Cumulative transactions reviewed	Cumulative frauds captured	Cumulative frauds captured by random guess
9	0	1	0	0.5
6	0	2	0	1
2	0	3	0	1.5
1	1	4	1	2
4	1	5	2	2.5
8	0	6	2	3
3	1	7	3	3.5
10	1	8	4	4
7	0	9	4	4.5
5	1	10	5	5

We can plot the cumulative frauds captured by the random model and also the logistic regression model, as shown in Figure 3-4.

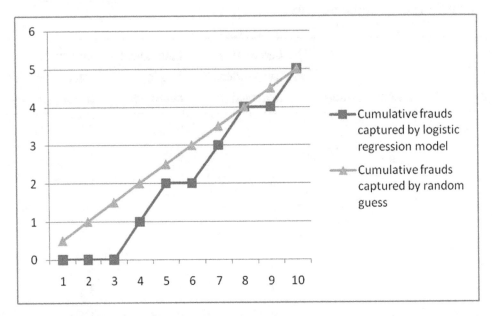

Figure 3-4. *Comparing the models*

In this particular case, for the example laid out above, random guess turned out to be *better* than the logistic regression model—in the first few guesses, a random guess makes better predictions than the model.

Now, let's get back to the previous fraudulent transaction example scenario, where let's say the outcome of model looks as follows:

	Cumulative frauds captured	
No. of transactions reviewed	Cumulative frauds captured by random guess	Cumulative frauds captued by model
-	-	0
100,000	1,000	4000
200,000	2,000	6000
300,000	3,000	7600
400,000	4,000	8100
500,000	5,000	8500

(*continued*)

	Cumulative frauds captured	
No. of transactions reviewed	Cumulative frauds captured by random guess	Cumulative frauds captued by model
600,000	6,000	8850
700,000	7,000	9150
800,000	8,000	9450
900,000	9,000	9750
1,000,000	10,000	10000

We lay out the chart between cumulative frauds captured by random guess versus the cumulative frauds captured by the model, as shown in Figure 3-5.

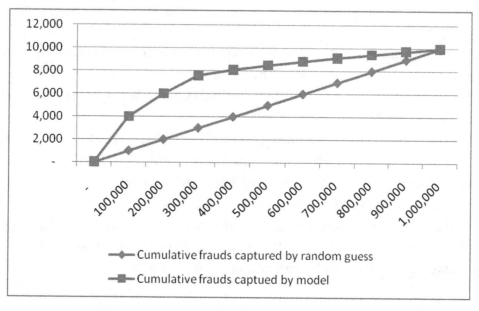

Figure 3-5. *Comparing the two approaches*

Note that the higher the area between the random guess line and the model line, the better the model performance is. The metric that measures the area covered under model line is called the *area under the curve* (AUC).

Thus, the AUC metric is a better metric to helps us evaluate the performance of a logistic regression model.

In practice, the output of rare event modeling looks as follows, when the scored dataset is divided into ten buckets (groups) based on probability (The code is provided as "credit default prediction.ipynb" in github):

	prediction avg_default	total_observations	SeriousDlqin2yrs avg_default	sum
prediction_rank				
1	0.005497	2250.0	0.003111	7
2	0.007639	2250.0	0.005333	12
3	0.009819	2250.0	0.007556	17
4	0.012689	2250.0	0.011556	26
5	0.017062	2250.0	0.015556	35
6	0.024761	2250.0	0.025778	58
7	0.038529	2250.0	0.042222	95
8	0.063098	2250.0	0.070222	158
9	0.120814	2250.0	0.123111	277
10	0.374364	2250.0	0.371556	836

prediction_rank in the preceding table represents the *decile* of probability—that is, after each transaction is rank ordered by probability and then grouped into buckets based on the decile it belongs to. Note that the third column (total_observations) has an equal number of observations in each decile.

The second column—prediction avg_default—represents the average probability of default obtained by the model we built. The fourth column—SeriousDlqin2yrs avg_default—represents the average actual default in each bucket. And the final column represents the actual number of defaults captured in each bucket.

Note that in an ideal scenario, all the defaults should be captured in the highest-probability buckets. Also note that, in the preceding table, the model captured considerable number of frauds in the highest probability bucket.

Common Pitfalls

This section talks about some the common pitfalls the analyst should be careful about while building a classification model:

Time Between Prediction and the Event Happening

Let's look at a case study: predicting the default of a customer.

We should say that it is useless to predict today that someone is likely to default on their credit card *tomorrow*. There should be some time gap between the time of predicting that someone would default and the event actually happening. The reason is that the operations team would take some time to intervene and help reduce the number of default transactions.

Outliers in Independent variables

Similar to how outliers in the independent variables impact the overall error in linear regression, it is better to cap outliers so that they do not impact the regression very much in logistic regression. Note that, unlike linear regression, logistic regression would not have a huge outlier output when one has an outlier input; in logistic regression, the output is always restricted between 0 and 1 and the corresponding cross entropy loss associated with it.

But the problem with having outliers would still result in a high cross entropy loss, and so it's a better idea to cap outliers.

Summary

In this chapter, we went through the following:

- Logistic regression is used in predicting binary (categorical) events, and linear regression is used to forecast continuous events.

- Logistic regression is an extension of linear regression, where the linear equation is passed through a sigmoid activation function.

- One of the major loss metrics used in logistic regression is the cross entropy error.

- A sigmoid curve helps in bounding the output of a value between 0 to 1 and thus in estimating the probability associated with an event.

- AUC metric is a better measure of evaluating a logistic regression model.

CHAPTER 4

Decision Tree

In the previous chapters, we've considered regression-based algorithms that optimize for a certain metric by varying coefficients or weights. A decision tree forms the basis of tree-based algorithms that help identify the rules to classify or forecast an event or variable we are interested in. Moreover, unlike linear or logistic regression, which are optimized for either regression or classification, decision trees are able to perform both.

The primary advantage of decision trees comes from the fact that they are *business user friendly*—that is, the output of a decision tree is intuitive and easily explainable to the business user.

In this chapter we will learn the following:

- How a decision tree works in classification and regression exercise

- How a decision tree works when the independent variable is continuous or discrete

- Various techniques involved in coming up with an optimal decision tree

- Impact of various hyper-parameters on a decision tree

- How to implement a decision tree in Excel, Python, and R

A *decision tree* is an algorithm that helps in classifying an event or predicting the output values of a variable. You can visualize decision trees as a set of rules based on which a different outcome can be expected. For example, look at Figure 4-1.

© V Kishore Ayyadevara 2018

V. K. Ayyadevara, *Pro Machine Learning Algorithms*, https://doi.org/10.1007/978-1-4842-3564-5_4

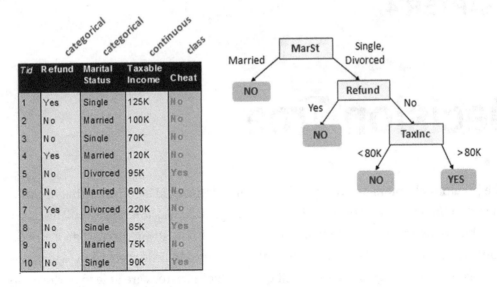

Figure 4-1. *An example decision tree*

In Figure 4-1, we can see that a dataset (the table on the left) uses both a continuous variable (Taxable Income) and categorical variables (Refund, Marital Status) as independent variables to classify whether someone was cheating on their taxes or not (categorical dependent variable).

The tree on the right has a few components: root node, decision nodes, and leaf node (I talk more about these in next section) to classify whether someone would cheat (Yes/No).

From the tree shown, the user can derive the following rules:

1. Someone with Marital Status of Yes is generally not a cheater.

2. Someone who is divorced but also got a refund earlier also does not cheat.

3. Someone who is divorced, did not get a refund, but has a taxable income of less than 80K is also not a cheater.

4. Those who do not belong to any of the preceding categories are cheaters in this particular dataset.

Similar to regression, where we derived an equation (for example, to predict credit default based on customer characteristics) a decision tree also works to predict or forecast an event based on customer characteristics (for example, marital status, refund, and taxable income in the previous example).

When a new customer applies for a credit card, the *rules engine* (a decision tree running on the back end) would check whether the customer would fall in the risky bucket or the non-risky bucket after passing through all the rules of the decision tree. After passing through the rules, the system would approve or deny the credit card based on the bucket a user falls into.

Obvious advantages of decision trees are the intuitive output and visualization that help a business user make a decision. Decision trees are also less sensitive to outliers in cases of classification than a typical regression technique. Moreover, a decision tree is one of the simpler algorithms in terms of building a model, interpreting a model, or even implementing a model.

Components of a Decision Tree

All the components of a decision tree are shown in Figure 4-2.

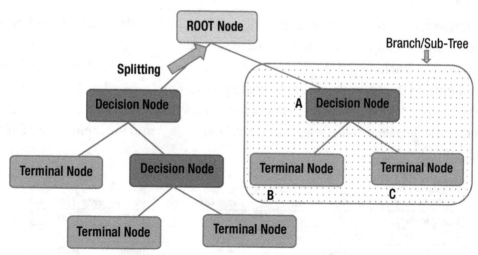

Note:- A is parent node of B and C.

Figure 4-2. Components of a decision tree

The components include the following:

- *Root node*: This node represents an entire population or sample and gets divided into two or more homogeneous sets.

- *Splitting*: A process of dividing a node into two or more sub-nodes based on a certain rule.

- *Decision node*: When a sub-node splits into further sub-nodes, it is called a decision node.

- *Leaf/Terminal node*: The final node in a decision tree.

- *Pruning*: The process of removing sub-nodes from a decision node—the opposite of splitting.

- *Branch/Sub-tree*: A subsection of an entire tree is called a branch or sub-tree.

- *Parent and child node*: A node that is divided into sub-nodes is called the parent node of those sub-nodes, and the sub-nodes are the children of the parent node.

Classification Decision Tree When There Are Multiple Discrete Independent Variables

The criterion for splitting at a root node varies by the type of variable we are predicting, depending on whether the dependent variable is continuous or categorical. In this section, we'll look at how splitting happens from the root node to decision nodes by way of an example. In the example, we are trying to predict employee salary (emp_sal) based on a few independent variables (education, marital status, race, and sex).

Here is the dataset (available as "categorical dependent and independent variables. xlsx" in github):

	A	B	C	D	E
	Edu_of_Emp	marital_Status	Emp_race_type	sex_of_emp	Emp_Sal
1					
2	Bachelors	Never-married	White	Male	<=50K
3	Bachelors	Married-civ-spouse	White	Male	<=50K
4	HS-grad	Divorced	White	Male	<=50K
5	11th	Married-civ-spouse	Black	Male	<=50K
6	Bachelors	Married-civ-spouse	Black	Female	<=50K
7	Masters	Married-civ-spouse	White	Female	<=50K
8	9th	Married-spouse-absent	Black	Female	<=50K
9	HS-grad	Married-civ-spouse	White	Male	>50K
10	Masters	Never-married	White	Female	>50K
11	Bachelors	Married-civ-spouse	White	Male	>50K
12	Some-college	Married-civ-spouse	Black	Male	>50K
13	Bachelors	Married-civ-spouse	Asian-Pac-Islander	Male	>50K
14	Bachelors	Never-married	White	Female	<=50K
15	Assoc-acdm	Never-married	Black	Male	<=50K

Here, `Emp_sal` is the dependent variable, and the rest of the variables are independent variables.

When splitting the root node (the original dataset), you first need to determine the variable based on which the first split has to be made—for example, whether you will split based on education, marital status, race, or sex. To come up with a way of shortlisting one independent variable over the rest, we use the information gain criterion.

Information Gain

Information gain is best understood by relating it to *uncertainty*. Let's assume that there are two parties contesting in elections being conducted in two different states. In one state the chance of a win is 50:50 for each party, whereas in another state the chance of a win for party A is 90% and for party B it's 10%.

If we were to predict the outcome of elections, the latter state is much easier to predict than the former because uncertainty is the least (the probability of a win for party A is 90%) in that state. Thus, information gain is a measure of uncertainty after splitting a node.

Calculating Uncertainty: Entropy

Uncertainty, also called *entropy*, is measured by the formula

$$-\left(plog_2p+qlog_2q\right)$$

where p is the probability of event 1 happening, and q is the probability of event 2 happening.

Let's consider the win scenarios for the two parties:

Scenario	Party A uncertainty	Party B uncertainty	Overall uncertainty
Equal chances of win	$-0.5log_2(0.5) = 0.5$	$-0.5log_2(0.5) = 0.5$	$0.5 + 0.5 = 1$
90% chances of win for party A	$-0.9log_2(0.9) = 0.1368$	$-0.1log_2(0.1) = 0.3321$	$0.1368 + 0.3321 = 0.47$

We see that based on the previous equation, the second scenario has less overall uncertainty than the first, because the second scenario has a 90% chance of party A's win.

Calculating Information Gain

We can visualize the root node as the place where maximum uncertainty exists. As we intelligently split further, the uncertainty decreases. Thus, the choice of split (the variable the split should be based on) depends on which variables decrease uncertainty the most.

To see how the calculation happens, let's build a decision tree based on our dataset.

Uncertainty in the Original Dataset

In the original dataset, nine observations have a salary <= 50K, while five have a salary > 50K:

<=50K	>50K	Grand Total
9	5	14

Let's calculate the values of p and q so that we calculate the overall uncertainty:

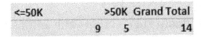

	B	C
4	<=50K	>50K
5	9	5
6		
7	p	q
8	0.64	0.36

The formulas for p and q are as follows:

	B	C
4	<=50K	>50K
5	9	5
6		
7	p	q
8	=B5/D5	=C5/D5

Thus, the overall uncertainty in the root node is as follows:

Uncertainty in <=50K	Uncertainty in >50K	Overall uncertainty
$-0.64 \times \log_2(0.64) = 0.41$	$-0.36 \times \log_2(0.36) = 0.53$	$0.41 + 0.53 = 0.94$

The overall uncertainty in the root node is 0.94.

To see the process of shortlisting variables in order to complete the first step, we'll figure out the amount by which overall uncertainty decreases if we consider all four independent variables for the first split. We'll consider education for the first split (we'll figure out the improvement in uncertainty), next we'll consider marital status the same way, then race, and finally the sex of the employee. The variable that reduces uncertainty the most will be the variable we should use for the first split.

Measuring the Improvement in Uncertainty

To see how the improvement in uncertainty is calculated, consider the following example. Let's consider whether we want to split our variable by sex of employee:

	A	B	C	D
1				
2				
3	Count of Edu_of_Emp	Column Labels ▾		
4	Row Labels ▾	<=50K	>50K	Grand Total
5	Female		4	1	5
6	Male		5	4	9
7	Grand Total		9	5	14

We calculate the uncertainty $-(p\log_2 p + q\log_2 q)$ of each distinct value of each variable. The table for the uncertainty calculation for one of the variables (Sex) is as follows:

Sex	P	Q	$-(p\log_2 p)$	$-(q\log_2 q)$	$-(p\log_2 p + q\log_2 q)$	Weighted uncertainty
Female	4/5	1/5	0.257	0.46	0.72	$0.72 \times 5/14 = 0.257$
Male	5/9	4/9	0.471	0.52	0.99	$0.99 \times 9/14 = 0.637$
Overall						**0.894**

A similar calculation to measure the overall uncertainty of all the variables would be done. The information gain if we split the root node by the variable Sex is as follows:

(Original entropy – Entropy if we split by the variable Sex) = 0.94 – 0.894 = 0.046

Based on the overall uncertainty, the variable that maximizes the information gain (the reduction in uncertainty) would be chosen for splitting the tree.

In our example, variable-wise, overall uncertainty is as follows:

Variable	Overall uncertainty	Reduction in uncertainty from root node
Education	0.679	0.94 – 0.679 = 0.261
Marital status	0.803	0.94 – 0.803 = 0.137
Race	0.803	0.94 – 0.803 = 0.137
Sex	0.894	0.94 – 0.894 = 0.046

From that we can observe that the splitting decision should be based on Education and not any other variable, because it is the variable that reduces the overall uncertainty by the most (from 0.94 to 0.679).

Once a decision about the split has been made, the next step (in the case of variables that have more than two distinct values—education in our example) would be to determine which unique value should go to the right decision node and which unique value should go to the left decision node after the root node.

Let's look at all the distinct values of education, because it's the variable that reduces uncertainty the most:

Distinct value	% of obs. <=50K
11th	100%
9th	100%
Assoc-acdm	100%
Bachelors	67%
HS-grad	50%
Masters	50%
Some-college	0%
Overall	**64%**

Which Distinct Values Go to the Left and Right Nodes

In the preceding section, we concluded that education is the variable on which the first split in the tree would be made. The next decision to be made is which distinct values of education go to the left node and which distinct values go to the right node.

The Gini impurity metric comes in handy in such scenario.

Gini Impurity

Gini impurity refers to the extent of inequality within a node. If a node has all values that belong to one class over the other, it is the *purest* possible node. If a node has 50% observations of one class and the rest of another class, it is the most *impure* form of a node.

Gini impurity is defined as $1-\left(p^2 + q^2\right)$ where p and q are the probabilities associated with each class.

Consider the following scenario:

P	Q	Gini index value
0	1	$1 - 0^2 - 1^2 = 0$
1	0	$1 - 1^2 - 0^2 = 0$
0.5	0.5	$1 - 0.5^2 - 0.5^2 = 0.5$

Let's use Gini impurity for the employee salary prediction problem: From the information gain calculation in the previous section, we observed that Education is the variable used as the first split.

To figure out which distinct values go to the left node and which go to the right node, let's go through the following calculation:

	# of observations			Left node		Right node		Impurity		# of obs		Weighted impurity
	<=50K	>50K	Grand Total	p	q	p	q	Left node	Right node	Left node	Right node	
11th	1		1	100%	0%	62%	38%	-	0.47	1	13	0.44
9th	1		1	100%	0%	58%	42%	-	0.49	2	12	0.42
Assoc-acdm	1		1	100%	0%	55%	45%	-	0.50	3	11	0.39
Bachelors	4	2	6	78%	22%	40%	60%	0.35	0.48	9	5	0.39
HS-grad	1	1	2	73%	27%	33%	67%	0.40	0.44	11	3	0.41
Masters	1	1	2	69%	31%	0%	100%	0.43	-	13	1	0.40
Some-college	1	1										

We can understand the preceding table through the following steps:

1. Rank order the distinct values by the percentage of observations that belong to a certain class.

   This would result in the distinct values being reordered as follows:

Distinct value	% of obs <= 50K
11th	100%
9th	100%
Assoc-acdm	100%
Bachelors	67%
HS-grad	50%
Masters	50%
Some-college	0%

2. In the first step, let's assume that only the distinct values that correspond to 11th go to the left node, and the rest of the distinct observations correspond to the right node.

   Impurity in the left node is 0 because it has only one observation, and there will be some impurity in the right node because eight observations belong to one class and five belong to another class.

3. Overall impurity is calculated as follows:

   ((Impurity in left node × # of obs. in left node) + (Impurity in right node × # of obs. in right node)) / (total no. of obs.)

4. We repeat steps 2 and 3 but this time we include both 11th and 9th in the left node and the rest of the distinct values in the right node.

5. The process is repeated until all the distinct values are considered in the left node.

6. The combination that has the least overall weighted impurity is the combination that will be chosen in the left node and right node.

In our example, the combination of {11th,9th,Assoc-acdm} goes to the left node and the rest go to the right node as this combination has the least weighted impurity.

Splitting Sub-nodes Further

From the analysis so far, we have split our original data into the following:

3	Count of Edu_of_Emp	Column Labels		
4	Row Labels	<=50K	>50K	Grand Total
5	Left node	3		3
6	Right node	6	5	11
7	Grand Total	9	5	14

Now there is an opportunity to split the right node further. Let us look into how the next split decision is. The data that is left to split is all the data points that belong to the right node, which are as follows:

Edu_of_Emp	marital_Status	Emp_race_type	sex_of_emp	Emp_Sal
Bachelors	Never-married	White	Male	<=50K
Bachelors	Married-civ-spouse	White	Male	<=50K
HS-grad	Divorced	White	Male	<=50K
Bachelors	Never-married	White	Female	<=50K
Bachelors	Married-civ-spouse	Black	Female	<=50K
Masters	Married-civ-spouse	White	Female	<=50K
Bachelors	Married-civ-spouse	Asian-Pac-Islander	Male	>50K
HS-grad	Married-civ-spouse	White	Male	>50K
Masters	Never-married	White	Female	>50K
Bachelors	Married-civ-spouse	White	Male	>50K
Some-college	Married-civ-spouse	Black	Male	>50K

From the preceding data, we'll perform the following steps:

1. Use the information gain metric to identify the variable that should be used to split the data.

2. Use the Gini index to figure out the distinct values within that variable that should belong to the left node and the ones that should belong to the right node.

The overall impurity in the parent node for the preceding dataset is as follows:

<=50K	>50K	Grand Total
6	5	11

<=50K	>50K
6	5

p	q
0.55	0.45

plogp	qlogq
-0.476983155	-0.517

Overall impurity	0.994

Now that the overall impurity is ~0.99, let's look at the variable that reduces the overall impurity by the most. The steps we would go through would be the same as in the previous iteration on the overall dataset. Note that the only difference between the current version and the previous one is that in the root node the total dataset is considered, whereas in the sub-node only a subset of the data is considered.

Let's calculate the information gain obtained by each variable separately (similar to the way we calculated in the previous section). The overall impurity calculation with marital status as the variable would be as follows:

	Row Labels	<=50K	>50K	Grand Total	p	q	plogp	qlogp	-(plogp+qlogq)
4									
5	Divorced	1		1	1.00	-	-	-	-
6	Married-civ-spouse	3	4	7	0.43	0.57	(0.52)	(0.46)	0.99
7	Never-married	2	1	3	0.67	0.33	(0.39)	(0.53)	0.92
8	Grand Total	6	5	11			Overall impurity		0.88

In a similar manner, impurity with respect to employee race would be as follows:

	Row Labels	<=50K	>50K	Grand Total	p	q	plogp	qlogp	-(plogp+qlogq)
4									
5	Asian-Pac-Islander		1	1	-	1.00	-	-	-
6	Black	1	1	2	0.50	0.50	(0.50)	(0.50)	1.00
7	White	5	3	8	0.63	0.38	(0.42)	(0.53)	0.95
8	Grand Total	6	5	11			Overall impurity		0.88

Impurity with respect to sex would be calculated as follows:

	Row Labels	<=50K	>50K	Grand Total	p	q	plogp	qlogp	-(plogp+qlogq)
4									
5	Female	3	1	4	0.75	0.25	(0.31)	(0.50)	0.81
6	Male	3	4	7	0.43	0.57	(0.52)	(0.46)	0.99
7	Grand Total	6	5	11			Overall impurity		0.92

Impurity with respect to employee education would be calculated as follows:

	Row Labels	<=50K	>50K	Grand Total	p	q	plogp	qlogp	-(plogp+qlogq)
4									
5	Bachelors	4	2	6	0.67	0.33	(0.39)	(0.53)	0.92
6	HS-grad	1	1	2	0.50	0.50	(0.50)	(0.50)	1.00
7	Masters	1	1	2	0.50	0.50	(0.50)	(0.50)	1.00
8	Some-college		1	1	-	1.00	-	-	-
9	Grand Total	6	5	11			Overall impurity		0.86

From the above, we notice that employee education as a variable is the one that reduces impurity the most from the parent node—that is, the highest information gain is obtained from the employee education variable.

Notice that, coincidentally, the same variable happened to split the dataset twice, both in the parent node and in the sub-node. This pattern might not be repeated on a different dataset.

When Does the Splitting Process Stop?

Theoretically, the process of splitting can happen until all the terminal (leaf/last) nodes of a decision tree are pure (they all belong to one class or the other).

However, the disadvantage of such process is that it overfits the data and hence might not be generalizable. Thus, a decision tree is a trade-off between the complexity of the tree (the number of terminal nodes in a tree) and its accuracy. With a lot of terminal nodes, the accuracy might be high on training data, but the accuracy on validation data might not be high.

This brings us to the concept of the complexity parameter of a tree and out-of-bag validation. As the complexity of a tree increases—that is, the tree's depth gets higher—the accuracy on the training dataset would keep increasing, but the accuracy on the test dataset might start getting lower beyond certain depth of the tree.

The splitting process should stop at the point where the validation dataset accuracy does not improve any further.

Classification Decision Tree for Continuous Independent Variables

So far, we have considered that both the independent and dependent variables are categorical. But in practice we might be working on continuous variables too, as independent variables. This section talks about how to build a decision tree for a continuous variable.

We will look at the ways in which we can build a decision tree for a categorical dependent variable and continuous independent variables in the next sections using the following dataset ("categorical dependent continuous independent variable.xlsx" in github):

Survived	Age
1	16
1	7
1	15
1	10
1	3
1	20
0	28
0	94
0	62
0	76
0	34
0	26

In this dataset, we'll try to predict whether someone would survive or not based on the Age variable.

The dependent variable is Survived, and the independent variable is Age.

1. Sort the dataset by increasing independent variable. The dataset thus transforms into the following:

Survived	Age
1	3
1	7
1	10
1	15
1	16
1	20
0	26
0	28
0	34
0	62
0	76
0	94

2. Test out multiple rules. For example, we can test the impurity in both left and right nodes when Age is less than 7, less than 10 and so on until Age less than 94.

3. Calculate Gini impurity:

Survived	Count of Survived2	# of obs Left node	# of obs Right node	Left node p	Left node q	Left node Impurity	Right node p	Right node q	Right node Impurity	Overall impuritiy
3	1	1	11							
7	1	2	10	100%	0%	-	40%	60%	0.48	0.44
10	1	3	9	100%	0%	-	33%	67%	0.44	0.36
15	1	4	8	100%	0%	-	25%	75%	0.38	0.27
16	1	5	7	100%	0%	-	14%	86%	0.24	0.16
20	1	6	6	100%	0%	-	0%	100%	-	-
26	1	7	5	86%	14%	0.24	0%	100%	-	0.13
28	1	8	4	75%	25%	0.38	0%	100%	-	0.24
34	1	9	3	67%	33%	0.44	0%	100%	-	0.32
62	1	10	2	60%	40%	0.48	0%	100%	-	0.39
76	1	11	1	55%	45%	0.50	0%	100%	-	0.45
94	1	12								

From the preceding table, we should notice that Gini impurity is the least when the independent variable value is less than 26.

Thus we will choose Age less than 26 as the rule that splits the original dataset.

Note that both Age > 20 as well as Age < 26 split the dataset to the same extent of error rate. In such a scenario, we need to come up with a way to choose the one rule between the two rules. We'll take the average of both rules, so Age <= 23 would be the rule that is midway between the two rules and hence a better rule than either of the two.

Classification Decision Tree When There Are Multiple Independent Variables

To see how a decision tree works when multiple independent variables are continuous, let's go through the following dataset ("categorical dependent multiple continuous independent variables.xlsx" in github):

Survived	Age	Unknown
1	30	79
1	7	67
1	100	53
1	15	33
1	16	32
1	20	5
0	26	14
0	28	16
0	34	70
0	62	35
0	76	66
0	94	22

So far, we have performed calculations in the following order:

1. Identify the variable that should be used for the split first by using information gain.

2. Once a variable is identified, in the case of discrete variables, identify the unique values that should belong to the left node and the right node.

3. In the case of continuous variables, test out all the rules and shortlist the rule that results in minimal overall impurity.

In this case, we'll reverse the scenario—that is, we'll first find out the rule of splitting to the left node and right node, if we were to split on any of the variables. Once the left and right nodes are figured, we'll calculate the information gain obtained by the split and thus shortlist the variable that should be slitting the overall dataset.

First, we'll calculate the optimal split for both variables. We'll start with Age as the first variable:

Age	Survived
7	1
15	1
16	1
20	1
26	0
28	0
30	1
34	0
62	0
76	0
94	0
100	1

Age	Survived	Count of Survived2	# of obs Left node	# of obs Right node	Left node p	Left node q	Left node Impurity	Right node p	Right node q	Right node Impurity	Overall impurity
7	1	1		11							
15	1	1	2	10	100%	0%	-	40%	60%	0.48	0.40
16	1	1	3	9	100%	0%	-	33%	67%	0.44	0.33
20	1	1	4	8	100%	0%	-	25%	75%	0.38	0.25
26	0	1	5	7	80%	20%	0.32	29%	71%	0.41	0.37
28	0	1	6	6	67%	33%	0.44	33%	67%	0.44	0.44
30	1	1	7	5	71%	29%	0.41	20%	80%	0.32	0.37
34	0	1	8	4	63%	38%	0.47	25%	75%	0.38	0.44
62	0	1	9	3	56%	44%	0.49	33%	67%	0.44	0.48
76	0	1	10	2	50%	50%	0.50	50%	50%	0.50	0.50
94	0	1	11	1	45%	55%	0.50	100%	0%	-	0.45
100	1	1	12								

From the data, we can see that the rule derived should be Age <= 20 or Age >= 26. So again, we'll go with the middle value: Age <= 23.

Now that we've derived the rule, let's calculate the information gain corresponding to the split. Before calculating the information gain, we'll calculate the entropy in the original dataset:

	0	1	Grand Total
	6	6	12

Given that both 0 and 1 are 6 in number (50% probability for each), the overall entropy comes out to be 1.

From the following, we notice that entropy reduces to 0.54 from a value of 1 if we split the dataset first using the Age variable:

	Survived						
	0	1	Grand Total	p	q	-(plogp+qlogq)	wegihted entropy
Age<=23		4	4	0	1	0	0
Age>23	6	2	8	0.75	0.25	0.811278	0.540852
Grand Total	6	6	12		Overall entropy		0.540852

Similarly, had we split the dataset by the column named Unknown, the minima occurs when the value of Unknown <= 22, as follows:

Unknwon	Survived	Count of Survived2	# of obs Left node	# of obs Right node	Left node p	Left node q	Left node Impurity	Right node p	Right node q	Right node Impurity	Overall impurtiy
5	1	1	1	11							
14	0	1	2	10	50%	50%	0.50	50%	50%	0.50	0.50
16	0	1	3	9	33%	67%	0.44	56%	44%	0.49	0.48
22	0	1	4	8	25%	75%	0.38	63%	38%	0.47	0.44
32	1	1	5	7	40%	60%	0.48	57%	43%	0.49	0.49
33	1	1	6	6	50%	50%	0.50	50%	50%	0.50	0.50
35	0	1	7	5	43%	57%	0.49	60%	40%	0.48	0.49
53	1	1	8	4	50%	50%	0.50	50%	50%	0.50	0.50
66	0	1	9	3	44%	56%	0.49	67%	33%	0.44	0.48
67	1	1	10	2	50%	50%	0.50	50%	50%	0.50	0.50
70	0	1	11	1	45%	55%	0.50	100%	0%	-	0.45
79	1	1	12								

Thus, all values less than or equal to 22 belong to one group (left node), and the rest belong to another group (right node). Note that, practically we would go with a mid value between 22 and 32.

Overall entropy in case of splitting by the Unknown variable would be as follows:

	Survived						
	0	1	Grand Total	p	q	-(plogp+qlogq)	wegihted entropy
Unknown<=22	3	1	4	0.75	0.25	0.81	0.27
Unknown>22	3	5	8	0.375	0.625	0.95	0.64
Grand Total	6	6	12		Overall entropy		0.91

From the data, we see that information gain, due to splitting by the Unknown variable, is only 0.09. Hence, the split would be based on Age, not Unknown.

Classification Decision Tree When There Are Continuous and Discrete Independent Variables

We've seen ways of building classification decision tree when all independent variables are continuous and when all independent variables are discrete.

If some independent variables are continuous and the rest are discrete, the way we would build a decision tree is very similar to the way we built in the previous sections:

1. For the continuous independent variables, we calculate the optimal splitting point.

2. Once the optimal splitting point is calculated, we calculate the information gain associated with it.

3. For the discrete variable, we calculate the Gini impurity to figure out the grouping of distinct values within the respective independent variable.

4. Whichever is the variable that maximizes information gain is the variable that splits the decision tree first.

5. We continue with the preceding steps in further building the sub-nodes of the tree.

What If the Response Variable Is Continuous?

If the response variable is continuous, the steps we went through in building a decision tree in the previous section remain the same, except that instead of calculating Gini impurity or information gain, we calculate the squared error (similar to the way we minimized sum of squared error in regression techniques). The variable that reduces the overall mean squared error of the dataset will be the variable that splits the dataset.

To see how the decision tree works in the case of continuous dependent and independent variables, we'll go through the following dataset as an example (available as "continuous variable dependent and independent variables.xlsx" in github):

variable	response
-0.37535	1590
-0.37407	2309
-0.37341	815
-0.37316	2229
-0.37263	839
-0.37249	2295
-0.37248	1996

Here, the independent variable is named `variable`, and the dependent variable is named `response`. The first step would be to sort the dataset by the independent variable, as we did in the classification decision tree example.

Once the dataset is sorted by the independent variable of interest, our next step is to identify the rule that splits the dataset into left and right node. We might come up with multiple possible rules. The exercise we'll perform would be useful in shortlisting the one rule that splits the dataset most optimally.

variable	response	average response		squared error		
		left node	right node	left node	right node	Overall error
-0.37535	1590					
-0.37407	2309	1,590	1,747	-	2,603,561	2,603,561
-0.37341	815	1,950	1,635	258,481	2,224,773	2,483,253
-0.37316	2229	1,571	1,840	1,116,541	1,384,683	2,501,223
-0.37263	839	1,736	1,710	1,440,935	1,182,662	2,623,597
-0.37249	2295	1,556	2,146	2,084,263	44,701	2,128,964
-0.37248	1996					

	F	G	H	I	J	K	L
3			average response		squared error		
4	variable	respons	left node	right node	left node	right node	Overall error
5	-0.37534	1590					
6	-0.37406	2309	=AVERAGE($G$5:G5)	=AVERAGE(G6:$G$11)	=(G5-H6)^2	=(G6-I6)^2+(G7-I6)^2+(G8-I6)^2+	=SUM(J6:K6)
7	-0.37340	815	=AVERAGE($G$5:G6)	=AVERAGE(G7:$G$11)	=(G5-H7)^2+(G6-H7)^2	=(G7-I7)^2+(G8-I7)^2+(G9-I7)^2+	=SUM(J7:K7)
8	-0.37315	2229	=AVERAGE($G$5:G7)	=AVERAGE(G8:$G$11)	=(G5-H8)^2+(G6-H8)^2+(G7-H8)^	=(G8-I8)^2+(G9-I8)^2+(G10-I8)^2	=SUM(J8:K8)
9	-0.37262	839	=AVERAGE($G$5:G8)	=AVERAGE(G9:$G$11)	=(G5-H9)^2+(G6-H9)^2+(G7-H9)^	=(G9-I9)^2+(G10-I9)^2+(G11-I9)^	=SUM(J9:K9)
10	-0.37249	2295	=AVERAGE($G$5:G9)	=AVERAGE(G10:$G$11)	=(G5-H10)^2+(G6-H10)^2+(G7-H1	=(G10-I10)^2+(G11-I10)^2	=SUM(J10:K10)
11	-0.37247	1996					

From the preceding, we see that the minimal overall error occurs when `variable` < -0.37249. So, the points that belong to the left node will have an average response of 1,556, and the points that belong to the right node will have an average response of 2,146. Note that 1,556 is the average response of all the variable values that are less than the threshold that we derived earlier. Similarly, 2,146 is the average response of all the variable values that are greater than or equal to the threshold that we derived (0.37249).

Continuous Dependent Variable and Multiple Continuous Independent Variables

In classification, we considered information gain as a metric to decide the variable that should first split the original dataset. Similarly, in the case of multiple competing independent variables for a continuous variable prediction, we'll shortlist the variable that results in least overall error.

We'll add one additional variable to the dataset we previously considered:

var2	variable	response
0.84	-0.37535	1590
0.51	-0.37407	2309
0.75	-0.37341	815
0.44	-0.37316	2229
0.3	-0.37263	839
0.78	-0.37249	2295
0.1	-0.37248	1996

We already calculated the overall error for various possible rules of `variable` in the previous section. Let's calculate the overall error for the various possible rules of `var2`.

The first step is to sort the dataset by increasing value of `var2`. So, the dataset we will work on now transforms to the following:

var2	response
0.1	1996
0.3	839
0.44	2229
0.51	2309
0.75	815
0.78	2295
0.84	1590

The overall error calculations for various possible rules that can be developed using `var2` are as follows:

var2	response	average response		squared error		Overall error
		left node	right node	left node	right node	
0.1	1996					
0.3	839	1,996	1,680	-	2,538,872	2,538,872
0.44	2229	1,418	1,848	669,325	1,691,143	2,360,468
0.51	2309	1,688	1,752	1,108,346	1,509,311	2,617,657
0.75	815	1,843	1,567	1,397,577	1,096,017	2,493,593
0.78	2295	1,638	1,943	2,243,415	248,513	2,491,928
0.84	1590					

Note that overall error is the least when `var2 < 0.44`. However, when we compare the least overall error produced by `variable` and the least overall error produced by `var2`, `variable` produces the least overall error and so should be the variable that splits the dataset.

Continuous Dependent Variable and Discrete Independent Variable

To find out how it works to predict a continuous dependent variable using a discrete independent variable, we'll use the following dataset as an example, where "var" is the independent variable and "response" is the dependent variable:

var	response
a	1590
b	2309
c	815
a	2229
b	839
c	2295
a	1996

Let's pivot the dataset as follows:

Row Labels ▾	Average of response
a	1,938.33
b	1,574.00
c	1,555.00
Grand Total	1,724.71

We'll order the dataset by increasing average response value:

Row Labels ▾	Average of response
c	1,555.00
b	1,574.00
a	1,938.33
Grand Total	1,724.71

Now we'll calculate the optimal left node and right node combination. In the first scenario, only c would be in the left node and a, b will be in the right node. The average response in left node will be 1555, and the average response in right node will be the average of {1574,1938} = {1756}.

Overall error calculation in this scenario will look as follows:

var	response	prediction	squared error
a	1590	1,756	27,611
b	2309	1,756	305,625
c	815	1,555	547,600
a	2229	1,756	223,571
b	839	1,756	841,195
c	2295	1,555	547,600
a	1996	1,756	57,520
		Overall error	2,550,722

In the second scenario, we'll consider both {c,b} to belong to the left node and {a} to belong to the right node. In this case, the average response in left node will be the average of {1555,1574} = {1564.5}

The overall error in such case will be calculated as follows:

var	response	prediction	squared error
a	1590	1,938	121,336
b	2309	1,565	554,280
c	815	1,565	561,750
a	2229	1,938	84,487
b	839	1,565	526,350
c	2295	1,565	533,630
a	1996	1,938	3,325
		Overall error	2,385,160

We can see that the latter combination of left and right nodes yields the lesser overall error when compared to the previous combination. So, the ideal split in this case would be {b,c} belonging to one node and {a} belonging to another node.

Continuous Dependent Variable and Discrete, Continuous Independent Variables

In the case where there are multiple independent variables, where some variables are discrete and others are continuous, we follow the same steps as earlier:

1. Identify the optimal cut-off points for each variable individually.

2. Understand the variable that reduces uncertainty the most.

The steps to be followed remain the same as in the previous sections.

Implementing a Decision Tree in R

The implementation of classification is different from the implementation of regression (continuous variable prediction). Thus, a specification of the type of model is required as an input.

The following code snippet shows how we can implement a decision tree in R (available as "Decision tree.R" in github):

```
# import dataset
t=read.csv("D:/Pro ML book/Decision tree/dt_continuous_dep_indep.csv")
library(rpart)
# fit a decision tree using rpart function
fit=rpart(response~variable,method="anova", data=t
     ,control=rpart.control(minsplit=1,minbucket=2,maxdepth=2))
```

The decision tree is implemented using the functions available in a package named rpart. The function that helps in building a decision tree is also named rpart. Note that in rpart we specify the method parameter.

A method anova is used when the dependent variable is continuous, and a method class is used when the dependent variable is discrete.

You can also specify the additional parameters: minsplit, minbucket, and maxdepth (more on these in the next section).

Implementing a Decision Tree in Python

The Python implementation of a classification problem would make use of the DecisionTreeClassifier function in the sklearn package (The code is available as "Decision tree.ipynb" in github):

```
from sklearn.tree import DecisionTreeClassifier
depth_tree = DecisionTreeClassifier()
depth_tree.fit(X, y)
```

For a regression problem, the Python implementation would make use of the DecisionTreeRegressor function in the sklearn package:

```
from sklearn.tree import DecisionTreeRegressor
depth_tree = DecisionTreeRegressor()
depth_tree.fit(X, y)
```

Common Techniques in Tree Building

We saw earlier that the complexity parameter (the number of terminal nodes) could be one parameter for us to optimize for while checking the out-of-bag validation. Other common techniques used include the following:

- Restricting the number of observations in each terminal node to a minimum number (at least 20 observations in a node, for example)
- Specifying the maximum depth of a tree manually
- Specifying the minimum number of observations in a node for the algorithm to consider further splitting

We do all the above to avoid the tree overfitting upon our data. To understand the problem of overfitting, let's go through the following scenario:

1. The tree has a total of 90 depth levels (maxdepth = 90). In this scenario, the tree that gets constructed would have so many branches that it overfits on the training dataset but does not necessarily generalize on the test dataset.

2. Similar to the problem with maxdepth, minimum number of observations in a terminal node could also lead to overfitting. If we do not specify maxdepth and have a small number in the minimum number of observations in the terminal node parameter, the resulting tree again is likely to be huge, with multiple branches, and will again be likely to result in a overfitting to a training dataset and not generalizing for a test dataset.

3. Minimum number of observations in a node to further split is a parameter that is very similar to minimum observations in a node, except that this parameter restricts the number of observations in the parent node rather than the child node.

Visualizing a Tree Build

In R, one can plot the tree structure using the `plot` function available in the `rpart` package. The `plot` function plots the skeleton of the tree, and the `text` function writes the rules that are derived at various parts of the tree. Here's a sample implementation of visualizing a decision tree:

```
# import dataset
t=read.csv("D:/Pro ML book/Decision tree/dt_continuous_dep_discrete_
indep.csv")
library(rpart)
# fit a decision tree using rpart function
fit=rpart(response~variable,method="anova",data=t
      ,control=rpart.control(minsplit=1,minbucket=2,maxdepth=2))
plot(fit, margin=0.2)
text(fit, cex=.8)
```

The output of the preceding code looks like this:

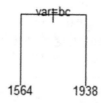

From this plot, we can deduce that when `var` is either of b or c, then the output is 1564. If not, the output is 1938.

In Python, one way to visualize a decision tree is by using a set of packages that have functions that help in display: Ipython.display, sklearn.externals.six, sklearn. tree, pydot, os.

```
from IPython.display import Image
from sklearn.externals.six import StringIO
from sklearn.tree import export_graphviz
import pydot
features = list(data.columns[1:])
```

```
import os
os.environ["PATH"] += os.pathsep + 'C:/Program Files (x86)/Graphviz2.38/bin/'
dot_data = StringIO()
export_graphviz(depth_tree, out_file=dot_data,feature_names=features,
filled=True,rounded=True)
graph = pydot.graph_from_dot_data(dot_data.getvalue())
Image(graph[0].create_png())
```

In the preceding code, you would have to change the data frame name in place of (data.columns[1:]) that we used in the first code snippet. Essentially, we are providing the independent variable names as features.

In the second code snippet, you would have to specify the folder location at which graphviz was installed and change the decision tree name to the name that the user has given in the fourth line (replace dtree with the variable name that you have created for DecisionTreeRegressor or DecisionTreeClassifier).

The output of the preceding code snippet looks like Figure 4-3.

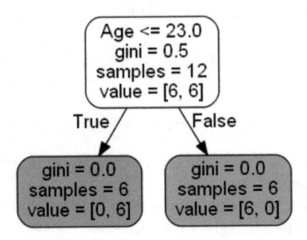

Figure 4-3. *The output of the code*

Impact of Outliers on Decision Trees

In previous chapters, we've seen that outliers have a big impact in the case of linear regression. However, in a decision tree outliers have little impact for classification, because we look at the multiple possible rules and shortlist the one that maximizes the information gain after sorting the variable of interest. Given that we are sorting the dataset by independent variable, there is no impact of outliers in the independent variable.

However, an outlier in the dependent variable in the case of a continuous variable prediction would be challenging if there is an outlier in the dataset. That's because we are using overall squared error as a metric to minimize. If the dependent variable contains an outlier, it causes similar issues like what we saw in linear regression.

Summary

Decision trees are simple to build and intuitive to understand. The prominent approaches used to build a decision tree are the combination of information gain and Gini impurity when the dependent variable is categorical and overall squared error when the dependent variable is continuous.

## CHAPTER 5

Random Forest

In Chapter 4, we looked at the process of building a decision tree. Decision trees can overfit on top of the data in some cases—for example, when there are outliers in dependent variable. Having correlated independent variables may also result in the incorrect variable being selected for splitting the root node.

Random forest overcomes those challenges by building multiple decision trees, where each decision tree works on a sample of the data. Let's break down the term: *random* refers to the random sampling of data from original dataset, and *forest* refers to the building of multiple decision trees, one each for each random sample of data. (Get it? It is a combination of multiple trees, and so is called a forest.)

In this chapter you will learn the following:

- Working details of random forest

- Advantages of random forest over decision tree

- The impact of various hyper-parameters on a decision tree

- How to implement random forest in Python and R

A Random Forest Scenario

To illustrate why random forest is an improvement over decision tree, let's go through a scenario in which we try to classify whether you will like a movie or not:

1. You ask for recommendations from a person A.

    a. A asks some questions to find out what your preferences are.

       i. We'll assume only 20 exhaustive questions are available.

       ii. We'll add the constraint that any person can ask from a random selection of 10 questions out of the 20 questions.

V. K. Ayyadevara, *Pro Machine Learning Algorithms*, https://doi.org/10.1007/978-1-4842-3564-5_5

      iii.   Given the 10 questions, the person ideally orders those questions in such a way that they are in a position to extract the maximum information from you.

    b.   Based on your responses, A comes up with a set of recommendations for movies.

2.   You ask for recommendations from another person, B.

    a.   As before, B asks questions to find out your preferences.

      i.   B is also in a position to ask only 10 questions out of the exhaustive list of 20 questions.

      ii.   Based on the set of 10 randomly selected questions, B again orders them to maximize the information obtained from your preferences.

      iii.   Note that the set of questions is likely to be different between A and B, though with some overlap.

    b.   Based on your responses, B comes up with recommendations.

3.   In order to achieve some randomness, you may have told A that *The Godfather* is the best movie you have ever watched. But you merely told B you enjoyed watching *The Godfather* "a lot."

    a.   This way, although the original information did not change, the way in which the two different people learned it was different.

4.   You perform the same experiment with n friends of yours.

    a.   Through the preceding, you have essentially built an *ensemble* of decision trees (where *ensemble* means *combination* of trees or forest).

5.   The final recommendation will be the average recommendation of all n friends.

Bagging

Bagging is short for *bootstrap aggregating*. The term *bootstrap* refers to selecting a few rows randomly (a sample from the original dataset), and *aggregating* means taking the average of predictions from all the decision trees that are built on the sample of the dataset.

This way, the predictions are less likely to be biased due to a few outlier cases (because there could be some trees built using sample data—where the sample data did not have any outliers). Random forest adopts the bagging approach in building predictions.

Working Details of a Random Forest

The algorithm for building a random forest is as follows:

1. Subset the original data so that the decision tree is built on only a sample of the original dataset.

2. Subset the independent variables (features) too while building the decision tree.

3. Build a decision tree based on the subset data where the subset of rows as well as columns is used as the dataset.

4. Predict on the test or validation dataset.

5. Repeat steps 1 through 3 *n* number of times, where *n* is the number of trees built.

6. The final prediction on the test dataset is the average of predictions of all *n* trees.

The following code in R builds the preceding algorithm (available as "rf_code.R" in github):

```
t=read.csv("train_sample.csv")
```

The described dataset has 140 columns and 10,000 rows. The first 8,000 rows are used to train the model, and the rest are used for testing:

```
train=t[1:8000,]
test=t[8001:9999,]
```

Given that we are using decision trees to build our random forest, let's use the rpart package:

```
library(rpart)
```

Initialize a new column named prediction in the test dataset:

```
test$prediction=0
for(i in 1:1000){ # we are running 1000 times - i.e., 1000 decision trees
  y=0.5
  x=sample(1:nrow(t),round(nrow(t)*y))  # Sample 50% of total rows, as y is 0.5
t2=t[x, c(1,sample(139,5)+1)]     # Sample 5 columns randomly, leaving the
first column which is the dependent variable
dt=rpart(response~.,data=t2)   # Build a decision tree on the above subset
of data (sample rows and columns)
 pred1=(predict(dt,test))  # Predict based on the tree just built
 test$prediction=(test$prediction+pred1)  # Add predictions of all the
 iterations of previously built decision trees
 }
test$prediction = (test$prediction)/1000  # Final prediction of the value
is the average of predictions of all the iterations
```

Implementing a Random Forest in R

Random forest can be implemented in R using the randomForest package. In the following code snippet, we try predicting whether a person would survive or not in the Titanic dataset (the code is available as "rf_code2.R" in github).

For simplicity, we will not deal with missing values and only consider those rows that do not have missing values:

```
t=read.csv("D:/Pro ML book/Random forest/titanic_train.csv")
head(t)
t2=na.omit(t)
library(randomForest)
rf=randomForest(Survived~.,data=t2,ntree=10)
```

In that code, we build a random forest that has 10 trees to provide predictions. The output of that code snippet is shown here:

```
> rf=randomForest(Survived~.,data=t2,ntree=10)
Error in randomForest.default(m, y, ...) :
  Can not handle categorical predictors with more than 53 categories.
In addition: Warning message:
In randomForest.default(m, y, ...) :
  The response has five or fewer unique values.  Are you sure you want to do
regression?
```

Note that the error message above specifies two things:

- The number of distinct values in some categorical independent variables is high.

- Additionally, it assumes that we would have to specify regression instead of classification.

Let's look at why random forest might have given an error when the categorical variable had a high number of distinct values. Note that random forest is an implementation of multiple decision trees. In a decision tree, when there are more distinct values, the frequency count of a majority of the distinct values is very low. When frequency is low, the purity of the distinct value is likely to be high (or impurity is low). But this isn't reliable, because the number of data points is likely to be low (in a scenario where the number of distinct values of a variable is high). Hence, random forest does not run when the number of distinct values in categorical variable is high.

Consider the warning message in the output: "The response has five or fewer unique values. Are you sure you want to do regression?" Note that, the column named Survived is numeric in class. So, the algorithm by default assumed that it is a regression that needs to be performed.

To avoid this, you need to convert the dependent variable into a categorical or factor variable:

```
t=read.csv("D:/Pro ML book/Random forest/titanic_train.csv")
head(t)
t2=na.omit(t)
t2$Survived=as.factor(t2$Survived)
library(randomForest)
rf=randomForest(Survived~Pclass+Sex+Age+SibSp+Parch+Fare+Embarked
                ,data=t2,ntree=10)
```

Now we can expect the random forest predictions to be made.

If we contrast the output with decision tree, one of the major drawbacks is that the decision tree output can be visualized as a tree, but random forest output cannot be visualized as a tree, because it is a combination of multiple decision trees. One way to understand the variable importance is by looking at how much of the decrease in overall impurity is because of splitting by the different variables.

Variable importance can be calculated in R by using the function `importance`:

```
> importance(rf)
            MeanDecreaseGini
Pclass           33.482304
Sex              88.140101
Age              46.483852
SibSp            17.759594
Parch             9.735269
Fare             39.454729
Embarked          7.469889
```

Consider how `MeanDecreaseGini` for the variable Sex might be calculated:

```
library(rpart)
samp=sample(nrow(t2),600)
train=t2[samp,]
test=t2[-samp,]
dt=rpart(Survived~Sex,data=train)
pred=predict(dt,test)
table(pred[,1],test$Survived)
```

In the preceding code snippet, a sample of the original dataset is selected. That sample is considered as a training dataset, and the rest are considered as a test dataset.

A decision tree is built based on the train dataset. Predictions are made on the *out-of-bag* data—that is, the test dataset:

```
> table(pred[,1],test$Survived)

                        0  1
   0.264840182648402    6 36
   0.79002624671916    59 13
```

Let us calculate the entropy of the preceding output:

		p	q	plogp	qlogq	obs.	Entropy
test	Original	0.42982	0.570175	-0.5236	-0.46214	114	0.985744
test	left node	0.26484	0.73516	-0.50765	-0.32632	42	0.833963
	right node	0.79003	0.209974	-0.26863	-0.4728	72	0.741433
					Overall entropy		**0.775523**

From the table, we can see that overall entropy reduces from 0.9857 to 0.7755.

Similarly, let's consider the other extreme, the variable that is least important: Embarked:

		p	q	plogp	qlogq	obs.	Entropy
test	Original	0.42982	0.570175	-0.5236	-0.46214	114	0.985744
test	left node	0.35185	0.648148	-0.53023	-0.40548	24	0.935711
	right node	0.65244	0.347561	-0.40196	-0.52991	90	0.931871
					Overall entropy		**0.932679**

From the preceding table, we see that entropy reduced from 0.9857 to only 0.9326. So, there was a much lower reduction in entropy when compared to the variable Sex. That means Sex as a variable is more important than Embarked.

The variable importance plot can also be obtained by the function shown in Figure 5-1.

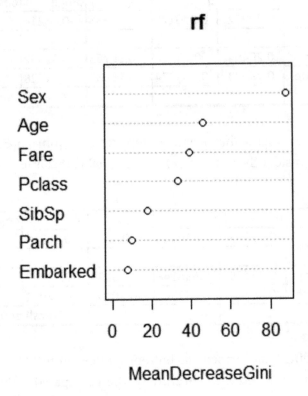

Figure 5-1. *Variable importance plot*

Parameters to Tune in a Random Forest

In the scenario just discussed, we noticed that random forest is based on decision tree, but on multiple trees running to produce an average prediction. Hence, the parameters that we tune in random forest would be very much the same as the parameters that are used to tune a decision tree.

The major parameters therefore are the following:

- Number of trees

- Depth of tree

To see the impact of number of trees on the test dataset AUC, we'll go through some code. AUC is calculated using the following code snippet:

```
library(ROCR)
pred_ROCR <- prediction(pred[,2], test$Survived)
auc_ROCR <- performance(pred_ROCR, measure = "auc")
auc_ROCR@y.values[[1]]
```

In the following code snippet, we increment the number of trees by a step of 10 and see how the AUC value varies over the number of trees:

```
auc=c()
no_of_trees=c()
for (i in 1:20){
  rf=randomForest(Survived~Pclass+Sex+Age+SibSp+Parch+Fare+Embarked
                  ,data=train,ntree=(10*i))
  pred=predict(rf,test,type="prob")
  pred_ROCR <- prediction(pred[,2], test$Survived)
  auc_ROCR <- performance(pred_ROCR, measure = "auc")
  auc=c(auc,auc_ROCR@y.values[[1]])
  no_of_trees=c(no_of_trees,(10*i))
}
plot(no_of_trees,auc)
```

In the first two lines, we have initialized empty vectors that we will keep populating over the for loop of different numbers of trees. After initialization, we are running a loop where the number of trees is incremented by 10 in each step. After running the random forest, we are calculating the AUC value of the predictions on the test dataset and keep appending the values of AUC.

The final plot of AUC over different numbers of trees is plotted in Figure 5-2.

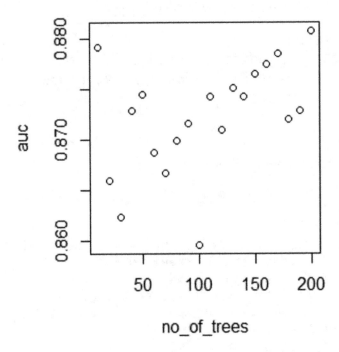

Figure 5-2. *AUC over different number of trees*

From Figure 5-2, we can see that as the number of trees increases, the AUC value of test dataset increases in general. But after a few more iterations, the AUC might not increase further.

Variation of AUC by Depth of Tree

In the previous section, we noted that the maximum value of AUC occurs when the number of trees is close to 200.

In this section, we'll consider the impact of depth of tree on the accuracy measure (AUC). In Chapter 4, we saw that the size of a node directly impacts the maximum depth of a tree. For example, if minimum possible node size is high, then depth is automatically low, and vice versa. So let's tune the node size as a parameter and see how AUC varies by node size:

```
auc=c()
node_size=c()
for (i in 1:100){
  rf=randomForest(Survived~Pclass+Sex+Age+SibSp+Parch+Fare+Embarked
                 ,data=train,nodesize=(5*i),ntree=200)
  pred=predict(rf,test,type="prob")
  pred_ROCR <- prediction(pred[,2], test$Survived)
  auc_ROCR <- performance(pred_ROCR, measure = "auc")
  auc=c(auc,auc_ROCR@y.values[[1]])
  node_size=c(node_size,(5*i))
}
plot(node_size,auc)
```

Note that the preceding code is very similar to the code we wrote for the variation in number of trees. The only addition is the parameter nodesize.

The output of the preceding code snippet is shown in Figure 5-3.

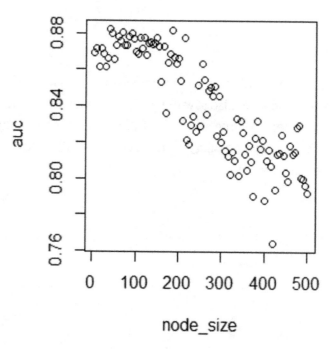

Figure 5-3. *AUC over different node sizes*

In Figure 5-3, note that as node size increases a lot, AUC of test dataset decreases.

Implementing a Random Forest in Python

Random forest is implemented in Python with the scikit-learn library. The implementation details of random forest are shown here (available in github as "random forest.ipynb"):

```
from sklearn.ensemble import RandomForestClassifier
rfc = RandomForestClassifier(n_estimators=100,max_depth=5,min_samples_
leaf=100,random_state=10)
rfc.fit(X_train, y_train)
```

Predictions are made as follows:

```
rfc_pred=rfc.predict_proba(X_test)
```

Once the predictions are made, AUC can be calculated as follows:

```
from sklearn.metrics import roc_auc_score
roc_auc_score(y_test, rfc_pred[:,1])
```

Summary

In this chapter, we saw how random forest improves upon decision tree by taking an average prediction approach. We also saw the major parameters that need to be tuned within a random forest: depth of tree and the number of trees. Essentially, random forest is a bagging (bootstrap aggregating) algorithm—it combines the output of multiple decision trees to give the prediction.

CHAPTER 6

Gradient Boosting Machine

So far, we've considered decision trees and random forest algorithms. We saw that random forest is a *bagging* (bootstrap aggregating) algorithm—it combines the output of multiple decision trees to give the prediction. Typically, in a bagging algorithm trees are grown in parallel to get the average prediction across all trees, where each tree is built on a sample of original data.

Gradient boosting, on the other hand, does the predictions using a different format. Instead of parallelizing the tree building process, boosting takes a sequential approach to obtaining predictions. In gradient boosting, each decision tree predicts the error of the previous decision tree—thereby *boosting* (improving) the error (gradient).

In this chapter, you will learn the following:

- Working details of gradient boosting

- How gradient boosting is different from random forest

- The working details of AdaBoost

- The impact of various hyper-parameters on boosting

- How to implement gradient boosting in R and Python

Gradient Boosting Machine

Gradient refers to the error, or *residual*, obtained after building a model. *Boosting* refers to improving. The technique is known as *gradient boosting machine*, or GBM. Gradient boosting is a way to gradually improve (reduce) error.

© V Kishore Ayyadevara 2018
V. K. Ayyadevara, *Pro Machine Learning Algorithms*, https://doi.org/10.1007/978-1-4842-3564-5_6

To see how GBM works, let's begin with an easy example. Assume you're given a model M (which is based on decision tree) to improve upon. Let's say the current model accuracy is 80%. We want to improve on that.

We'll express our model as follows:

```
Y = M(x) + error
```

Y is the dependent variable and *M(x)* is the decision tree using the *x* independent variables.

Now we'll predict the error from the previous decision tree:

```
error = G(x) + error2
```

G(x) is another decision tree that tries to predict the error using the *x* independent variables.

In the next step, similar to the previous step, we build a model that tries to predict error2 using the *x* independent variables:

```
error2 = H(x) + error3
```

Now we combine all these together:

```
Y = M(x) + G(x) + H(x) + error3
```

The preceding equation is likely to have an accuracy that is greater than 80% as individually model M (single decision tree) had 80% accuracy, while in the above equation we are considering 3 decision trees.

The next section explores the working details of how GBM works. In a later section, we will see how an *AdaBoost* (adaptive boosting) algorithm works.

Working details of GBM

Here's the algorithm for gradient boosting:

1. Initialize predictions with a simple decision tree.

2. Calculate residual - which is the (actual-prediction) value.

3. Build another shallow decision tree that predicts residual based on all the independent values.

4. Update the original prediction with the new prediction multiplied by learning rate.

5. Repeat steps 2 through 4 for a certain number of iterations (the number of iterations will be the number of trees).

Code implementing the preceding algorithm is as follows (code and dataset available in github as "GBM working details.ipynb"):

```
import pandas as pd
# importing dataset
data=pd.read_csv('F:/course/Logistic regression/credit_training.csv')
# removing irrelevant variables
data=data.drop(['Unnamed: 0'],axis=1)
# replacing null values
data['MonthlyIncome']=data['MonthlyIncome'].fillna(value=data
['MonthlyIncome'].median())
data['NumberOfDependents']=data['NumberOfDependents'].fillna(value=data
['NumberOfDependents'].median())
from sklearn.model_selection import train_test_split
# creating independent variables
X = data.drop('SeriousDlqin2yrs',axis=1)
# creating dependent variables
y = data['SeriousDlqin2yrs']
# creating train and test datasets
X_train, X_test, y_train, y_test = train_test_split(X, y, test_size=0.30,
random_state=42)
```

In the preceding code, we have split the dataset into 70% training and 30% test dataset.

```
# Build a decision tree
from sklearn.tree import DecisionTreeClassifier
depth_tree = DecisionTreeClassifier(criterion = "gini",max_depth=4,
min_samples_leaf=10)
depth_tree.fit(X_train, y_train)
```

In the preceding code, we are building a simple decision tree on the original data with SeriousDlqin2yrs as dependent variable and the rest of the variables as independent variables.

```
#Get the predictions on top of train dataset itself
dt_pred = depth_tree.predict_proba(X_train)
X_train['prediction']=dt_pred[:,1]
```

In the preceding code, we are predicting the outputs of the first decision tree. This will help us in coming up with the residuals.

```
#Get the predictions on top of test dataset
X_test['prediction']=depth_tree.predict_proba(X_test)[:,1]
```

In the preceding code, though we calculate the output probabilities in the test dataset, note that we are not in a position to calculate residuals because, as a practical matter, we are not allowed to peek into the dependent variable of the test dataset. As a continuation to the preceding code, we will be build 20 decision trees of residuals in the following code:

```
from sklearn.tree import DecisionTreeRegressor
import numpy as np
from sklearn.metrics import roc_auc_score
depth_tree2 = DecisionTreeRegressor(criterion = "mse",max_depth=4,
min_samples_leaf=10)
for i in range(20):
    # Calculate residual
    train_errorn=y_train-X_train['prediction']
    # remove prediction variable that got appended to independent variable
    earlier
    X_train2=X_train.drop(['prediction'],axis=1)
    X_test2=X_test.drop(['prediction'],axis=1)
```

In the preceding code, note that we are calculating the residual of the *n*th decision tree. We are dropping the prediction column from the X_train2 dataset because the prediction column cannot be one of the independent variables in the subsequent model that gets built in the next iteration of the for loop.

```
    # Build a decision tree to predict the residuals using independent
    variables
    dt2=depth_tree2.fit(X_train2, train_errorn)
```

```
# predict the residual
dt_pred_train_errorn = dt2.predict(X_train2)
```

In the preceding code, we are fitting the decision tree where the dependent variable is the residual and the independent variables are the original independent variables of the dataset.

Once the decision tree is fit, the next step is to predict the residual (which was the dependent variable):

```
# update the predictions based on predicted residuals
X_train['prediction']=(X_train['prediction']+dt_pred_train_errorn*1)
# Calculate AUC
train_auc=roc_auc_score(y_train,X_train['prediction'])
print("AUC on training data set is: "+str(train_auc))
```

In that code, the original predictions (which are stored in the X_train dataset) are updated with the predicted residuals we obtained in the previous step.

Note that we are updating our predictions by the prediction of residual (dt_pred_train_errorn). We have explicitly given a *1 in the preceding code because the concept of shrinkage or learning rate will be explained in the next section (the *1 will be replaced by *learning_rate).

Once the predictions are updated, we calculate the AUC on the training dataset:

```
# update the predictions based on predicted residuals for test dataset
dt_pred_test_errorn = dt2.predict(X_test2)
X_test['prediction']=(X_test['prediction']+dt_pred_test_errorn)
# Calculate AUC
test_auc=roc_auc_score(y_test,X_test['prediction'])
print("AUC on test data set is: "+str(test_auc))
```

Here we update the predictions on the test dataset. We do not know the residuals of test dataset, but we update predictions on the test dataset on the basis of the decision tree that got built to predict the residuals of the training dataset. Ideally, if the test dataset has no residuals, the predicted residuals should be close to 0, and if the original decision tree of the test dataset had some residuals, then the predicted residuals would be away from 0.

Once the predictions of the test dataset are updated, we print out the AUC of the test dataset.

Let's look at the output of the preceding code:

```
AUC on training data set is: 0.848050623873
AUC on test data set is: 0.846682999544
AUC on training data set is: 0.853994647817
AUC on test data set is: 0.85192324761
AUC on training data set is: 0.859648767973
AUC on test data set is: 0.856681743799
AUC on training data set is: 0.859446889291
AUC on test data set is: 0.856626792978
AUC on training data set is: 0.861012776958
AUC on test data set is: 0.856121639446
AUC on training data set is: 0.861480302384
AUC on test data set is: 0.855853840437
AUC on training data set is: 0.862978701454
AUC on test data set is: 0.856643395073
AUC on training data set is: 0.863623994208
AUC on test data set is: 0.856491053006
AUC on training data set is: 0.863943741279
AUC on test data set is: 0.856382618954
AUC on training data set is: 0.86496378239
AUC on test data set is: 0.856646258466
AUC on training data set is: 0.865035697503
AUC on test data set is: 0.854946352302
AUC on training data set is: 0.865532766858
AUC on test data set is: 0.856335153632
AUC on training data set is: 0.865925302458
AUC on test data set is: 0.855010651024
AUC on training data set is: 0.866548883759
AUC on test data set is: 0.854685265118
AUC on training data set is: 0.86685703723
AUC on test data set is: 0.854614126956
AUC on training data set is: 0.867480486612
AUC on test data set is: 0.853665458607
AUC on training data set is: 0.868237899156
AUC on test data set is: 0.85230809079
AUC on training data set is: 0.868531976496
AUC on test data set is: 0.851979051865
AUC on training data set is: 0.868999146757
AUC on test data set is: 0.852059793161
AUC on training data set is: 0.86917359293
AUC on test data set is: 0.851634196484
```

Note that the AUC of train dataset keeps on increasing with more trees. But the AUC of the test dataset decreases after a certain iteration.

Shrinkage

GBM is based on decision tree. So, just like Random Forest algorithm, the accuracy of GBM depends on the depth of trees considered, number of trees built and the minimum number of observations in a terminal node. Shrinkage is an additional parameter in GBM. Let's see what happens to the output of the train and test dataset AUC if we change the learning rate/shrinkage. We'll initialize the learning rate to be equal to 0.05 and run more trees:

```
from sklearn.model_selection import train_test_split
# creating independent variables
X = data.drop('SeriousDlqin2yrs',axis=1)
# creating dependent variables
y = data['SeriousDlqin2yrs']
# creating train and test datasets
X_train, X_test, y_train, y_test = train_test_split(X, y, test_size=0.30,
random_state=42)

from sklearn.tree import DecisionTreeClassifier
depth_tree = DecisionTreeClassifier(criterion = "gini",max_depth=4,
min_samples_leaf=10)
depth_tree.fit(X_train, y_train)

#Get the predictions on top of train and test datasets
dt_pred = depth_tree.predict_proba(X_train)
X_train['prediction']=dt_pred[:,1]
X_test['prediction']=depth_tree.predict_proba(X_test)[:,1]
from sklearn.tree import DecisionTreeRegressor
import numpy as np
from sklearn.metrics import roc_auc_score
depth_tree2 = DecisionTreeRegressor(criterion = "mse",max_depth=4,
min_samples_leaf=10)
learning_rate = 0.05
```

```
for i in range(20):
    # Calculate residual
    train_errorn=y_train-X_train['prediction']
    # remove prediction variable that got appended to independent variable
    earlier
    X_train2=X_train.drop(['prediction'],axis=1)
    X_test2=X_test.drop(['prediction'],axis=1)
    # Build a decision tree to predict the residuals using independent
    variables
    dt2=depth_tree2.fit(X_train2, train_errorn)
    # predict the residual
    dt_pred_train_errorn = dt2.predict(X_train2)
    # update the predictions based on predicted residuals
    X_train['prediction']=(X_train['prediction']+dt_pred_train_
    errorn*learning_rate)
    # Calculate AUC
    train_auc=roc_auc_score(y_train,X_train['prediction'])
    print("AUC on training data set is: "+str(train_auc))
    # update the predictions based on predicted residuals for test dataset
    dt_pred_test_errorn = dt2.predict(X_test2)
    X_test['prediction']=(X_test['prediction']+dt_pred_test_
    errorn*learning_rate)
    # Calculate AUC
    test_auc=roc_auc_score(y_test,X_test['prediction'])
    print("AUC on test data set is: "+str(test_auc))
```

The output of the preceding code is:

Here is the output of the first few trees:

```
AUC on training data set is: 0.834919832926
AUC on test data set is: 0.841572089467
AUC on training data set is: 0.83507772301
AUC on test data set is: 0.841509310179
AUC on training data set is: 0.835129216483
AUC on test data set is: 0.84156401098
AUC on training data set is: 0.851181471112
AUC on test data set is: 0.854494097151
AUC on training data set is: 0.851599769713
AUC on test data set is: 0.854700147423
AUC on training data set is: 0.850998154059
AUC on test data set is: 0.854503409398
AUC on training data set is: 0.851765152435
AUC on test data set is: 0.855041760727
AUC on training data set is: 0.853647845371
AUC on test data set is: 0.856149694228
AUC on training data set is: 0.853668334254
AUC on test data set is: 0.856128589996
```

And here is the output of the last few trees:

```
AUC on training data set is: 0.867592316056
AUC on test data set is: 0.865735946552
AUC on training data set is: 0.867612424811
AUC on test data set is: 0.865733165501
AUC on training data set is: 0.867669261648
AUC on test data set is: 0.865792064346
AUC on training data set is: 0.867720524457
AUC on test data set is: 0.865766428181
AUC on training data set is: 0.867784973217
AUC on test data set is: 0.865834936517
AUC on training data set is: 0.86782647483
AUC on test data set is: 0.865877706892
AUC on training data set is: 0.867898524015
AUC on test data set is: 0.865924891117
AUC on training data set is: 0.867910592573
AUC on test data set is: 0.865914267744
AUC on training data set is: 0.867927083994
AUC on test data set is: 0.865922932573
```

Unlike in the previous case, where `learning_rate` = 1, lower `learning_rate` resulted in the test dataset AUC increasing consistently along with the training dataset AUC.

AdaBoost

Before proceeding to the other methods of boosting, I want to draw a parallel from what we've seen in previous chapters. While calculating the error metric for logistic regression, we could have gone with the traditional squared error. But we moved to the entropy error, because it penalizes more for high amount of error.

In a similar manner, residual calculation can vary by the type of dependent variable. For a continuous dependent variable, residual calculation can be Gaussian ((actual – prediction) of dependent variable), whereas for a discrete variable, residual calculation could be different.

Theory of AdaBoost

AdaBoost is short for *adaptive boosting*. Here is the high-level algorithm:

1. Build a weak *learner* (decision tree technique in this case) using only a few independent variables.

    a. Note that while building the first weak learner, the weightage associated with each observation is the same.

2. Identify the observations that are classified incorrectly based on the weak learner.

3. Update the weights of the observations in such a way that the misclassifications in the previous weak learner are given more weightage and the correct classifications in the previous weak learner are given less weightage.

4. Assign a weightage for each weak learner based on the accuracy of its predictions.

5. The final prediction will be based on the weighted average prediction of multiple weak learners.

Adaptive potentially refers to the updating of weights of observations, depending on whether the previous classification was correct or incorrect. *Boosting* potentially refers to assigning the weightages to each weak learner.

Working Details of AdaBoost

Let's look at an example of AdaBoost:

1. Build a weak learner.

    Let's say the dataset is the first two columns in the following table (available as "adaboost.xlsx" in github):

X	Y	Left node			Right node			# of observations		Weighted
		p	q	impurity	p	q	impurity	left node	right node	impurity
1	1	1.00	-	-	0.44	0.56	0.49	1	9	4.44
2	1	1.00	-	-	0.38	0.63	0.47	2	8	3.75
3	1	1.00	-	-	0.29	0.71	0.41	3	7	2.86
4	1	1.00	-	-	0.17	0.83	0.28	4	6	1.67
5	0	0.80	0.20	0.32	0.20	0.80	0.32	5	5	3.20
6	0	0.67	0.33	0.44	0.25	0.75	0.38	6	4	4.17
7	0	0.57	0.43	0.49	0.33	0.67	0.44	7	3	4.76
8	1	0.63	0.38	0.47	-	1.00	-	8	2	3.75
9	0	0.56	0.44	0.49	-	1.00	-	9	1	4.44
10	0	0.50	0.50	0.50	-	1.00	-	10	0	5.00

    Once we have the dataset, we build a weak learner (decision tree) according to the steps laid out to the right in the preceding table. From the table we can see that $X <= 4$ is the optimal splitting criterion for this first decision tree.

2. Calculate the error metric (the reason for having "Changed Y" and "Yhat" as new columns in the below table will be explained after step 4):

X	Y	Changed Y	Weights	Prediction	Accurate?	Yhat	error
1	1	1	0.1	1	Yes	1	0
2	1	1	0.1	1	Yes	1	0
3	1	1	0.1	1	Yes	1	0
4	1	1	0.1	1	Yes	1	0
5	0	-1	0.1	0	Yes	-1	0
6	0	-1	0.1	0	Yes	-1	0
7	0	-1	0.1	0	Yes	-1	0
8	1	1	0.1	0	No	-1	0.1
9	0	-1	0.1	0	Yes	-1	0
10	0	-1	0.1	0	Yes	-1	0
						Overall error	0.1

The formulae used to obtain the preceding table are as follows:

	E	F	G	H	I	J	K	L
16								
17	X	Y	Changed Y	Weights	Prediction	Accurate?	Yhat	error
18	1	1	=IF(F18=1,1,-1)	0.1	1	Yes	=IF(I18=1,1,-1)	=H18*IF(I18=F18,0,1)
19	2	1	=IF(F19=1,1,-1)	0.1	1	Yes	=IF(I19=1,1,-1)	=H19*IF(I19=F19,0,1)
20	3	1	=IF(F20=1,1,-1)	0.1	1	Yes	=IF(I20=1,1,-1)	=H20*IF(I20=F20,0,1)
21	4	1	=IF(F21=1,1,-1)	0.1	1	Yes	=IF(I21=1,1,-1)	=H21*IF(I21=F21,0,1)
22	5	0	=IF(F22=1,1,-1)	0.1	0	Yes	=IF(I22=1,1,-1)	=H22*IF(I22=F22,0,1)
23	6	0	=IF(F23=1,1,-1)	0.1	0	Yes	=IF(I23=1,1,-1)	=H23*IF(I23=F23,0,1)
24	7	0	=IF(F24=1,1,-1)	0.1	0	Yes	=IF(I24=1,1,-1)	=H24*IF(I24=F24,0,1)
25	8	1	=IF(F25=1,1,-1)	0.1	0	No	=IF(I25=1,1,-1)	=H25*IF(I25=F25,0,1)
26	9	0	=IF(F26=1,1,-1)	0.1	0	Yes	=IF(I26=1,1,-1)	=H26*IF(I26=F26,0,1)
27	10	0	=IF(F27=1,1,-1)	0.1	0	Yes	=IF(I27=1,1,-1)	=H27*IF(I27=F27,0,1)
28							Overall error	=SUM(L18:L27)

3. Calculate the weightage that should be associated with the first weak learner:

$0.5 \times \log((1 - error)/error) = 0.5 \times \log(0.9 / 0.1) = 0.5 \times \log(9) = 0.477$

4. Update the weights associated with each observation in such a way that the previous weak learner's misclassifications have high weight and the correct classifications have low weight (essentially, we are tuning the weights associated with each observation in such a way that, in the new iteration, we try and make sure that the misclassifications are predicted more accurately):

X	Y	Changed Y	Weights	Prediction	Accurate?	Yhat	error	Updated weights
1	1	1	0.1	1	Yes	1	0	0.06
2	1	1	0.1	1	Yes	1	0	0.06
3	1	1	0.1	1	Yes	1	0	0.06
4	1	1	0.1	1	Yes	1	0	0.06
5	0	-1	0.1	0	Yes	-1	0	0.06
6	0	-1	0.1	0	Yes	-1	0	0.06
7	0	-1	0.1	0	Yes	-1	0	0.06
8	1	1	0.1	0	No	-1	0.1	0.16
9	0	-1	0.1	0	Yes	-1	0	0.06
10	0	-1	0.1	0	Yes	-1	0	0.06
					Overall error		0.1	

Note that the updated weights are calculated by the following formula:

$$originalweight * e^{(-weightage\ of\ learner*yhat*changed\ y)}$$

That formula should explain the need for changing the discrete values of y from {0,1} to {–1,1}. By changing 0 to –1, we are in a position to perform the multiplication better. Also note that in the preceding formula, weightage associated with the learner in general would more often than not be positive.

When yhat and changed_y are the same, the exponential part of formula would be a lower number (as the – weightage of learner × yhat × changed_y part of the formula would be negative, and an exponential of a negative is a small number).

When yhat and changed_y are different values, that's when the exponential would be a bigger number, and hence the updated weight would be more than the original weight.

5. We observe that the updated weights we obtained earlier do not sum up to 1. We update each weight in such a way that the sum of weights of all observations is equal to 1. Note that, the moment weights are introduced, we can consider this as a regression exercise.

Now that the weight for each observation is updated, we repeat the preceding steps until the weight of the misclassified observations increases so much that it is now correctly classified:

X	Y	Weight	Rule	Average Y left node	Average Y right node	1/0 Prediction left node	1/0 Prediction right node	error left node	error right node	overall error
1	1	0.09	X<= 1	1.00	0.44	1.00	-	-	0.48	0.48
2	1	0.09	X<= 2	1.00	0.38	1.00	-	-	0.40	0.40
3	1	0.09	X<= 3	1.00	0.29	1.00	-	-	0.31	0.31
4	1	0.09	X<= 4	1.00	0.17	1.00	-	-	0.22	0.22
5	0	0.09	X<= 5	0.80	0.20	1.00	-	0.09	0.22	0.31
6	0	0.09	X<= 6	0.67	0.25	1.00	-	0.17	0.22	0.40
7	0	0.09	X<= 7	0.57	0.33	1.00	-	0.26	0.22	0.48
8	1	0.22	X<= 8	0.63	-	1.00	-	0.26	-	0.26
9	0	0.09	X<= 9	0.56	-	1.00	-	0.34	-	0.34
10	0	0.09	X<= 10	0.50	-	1.00	-	0.43	-	0.43

Note that the weights in the third column in the preceding table are updated based on the formula we derived earlier post *normalization* (ensuring that the sum of weights is 1).

You should be able to see that the weight associated with misclassification (the eighth observation with the independent variable value of 8) is more than any other observation.

Note that although everything is similar to a typical decision tree till the prediction columns, error calculation gives emphasis to weights of observations. *error* in the left node is the summation of the weight of each observation that was misclassified in the left node and similarly for the right node.

overall error is the summation of error across both nodes (error in left node + error in right node).

In this instance, *overall error* is still the least at the fourth observation.

The updated weights based on the previous step are as follows:

X	Y	Weight	Rule	overall error	final prediction	error	Updated weight	Normalized weight
1	1	0.09	X<= 1	0.48	1	0	0.07	0.07
2	1	0.09	X<= 2	0.40	1	0	0.07	0.07
3	1	0.09	X<= 3	0.31	1	0	0.07	0.07
4	1	0.09	X<= 4	0.22	1	0	0.07	0.07
5	0	0.09	X<= 5	0.31	0	0	0.07	0.07
6	0	0.09	X<= 6	0.40	0	0	0.07	0.07
7	0	0.09	X<= 7	0.48	0	0	0.07	0.07
8	1	0.22	X<= 8	0.26	0	0.2239173	0.29	0.33
9	0	0.09	X<= 9	0.34	0	0	0.07	0.07
10	0	0.09	X<= 10	0.43	0	0	0.07	0.07
					Overall error	0.2239173		

Continue the process one more time:

X	Y	Weight	Rule	Average Y		1/0 Prediction		error		
				left node	right node	left node	right node	left node	right node	overall error
1	1	0.07	X<= 1	1.00	0.44	1.00	-	-	0.55	0.55
2	1	0.07	X<= 2	1.00	0.38	1.00	-	-	0.48	0.48
3	1	0.07	X<= 3	1.00	0.29	1.00	-	-	0.41	0.41
4	1	0.07	X<= 4	1.00	0.17	1.00	-	-	0.33	0.33
5	0	0.07	X<= 5	0.80	0.20	1.00	-	0.07	0.33	0.41
6	0	0.07	X<= 6	0.67	0.25	1.00	-	0.15	0.33	0.48
7	0	0.07	X<= 7	0.57	0.33	1.00	-	0.22	0.33	0.55
8	1	0.33	X<= 8	0.63	-	1.00	-	0.22	-	0.22
9	0	0.07	X<= 9	0.56	-	1.00	-	0.30	-	0.30
10	0	0.07	X<= 10	0.50	-	1.00	-	0.37	-	0.37

From the preceding table, note that *overall error* in this iteration is minimum at $X <= 8$, as the weight associated with the eighth observation is a lot more than other observations, and hence *overall error* came up as the least at the eighth observation this time. However, note that the weightage associated with the preceding tree would be low, because the accuracy of that tree is low when compared to the previous two trees.

6. Once all the predictions are made, the final prediction for an observation is calculated as the summation of the weightage associated with each weak learner multiplied by the probability output for each observation.

Additional Functionality for GBM

In the previous section, we saw how to hand construct GBM. In this section, we will look at other parameters that can be built in:

- *Row sampling*: In random forest, we saw that sampling a random selection of rows results in a more generalized and better model. In GBM, too, we can potentially sample rows to improve the model performance further.

- *Column sampling*: Similar to row sampling, some amount of overfitting can be avoided by sampling columns for each decision tree.

Both random forest and GBM techniques are based on decision tree. However, a random forest can be thought of as building multiple trees in parallel, where in the end we take the average of all the multiple trees as the final prediction. In a GBM, we build multiple trees, but in sequence, where each tree tries to predict the residual of its previous tree.

Implementing GBM in Python

GBM can be implemented in Python using the `scikit-learn` library as follows (code is available as "GBM.ipynb" in github):

```
from sklearn import ensemble
gb_tree = ensemble.GradientBoostingClassifier(loss='deviance',
learning_rate=0.05,n_estimators=100,min_samples_leaf=10,max_depth=13,
max_features=2,subsample=0.7,random_state=10)
gb_tree.fit(X_train, y_train)
```

Note that the key input parameters are a loss function (whether it is a normal residual approach or an AdaBoost-based approach), learning rate, number of trees, depth of each tree, column sampling, and row sampling.

Once a GBM is built, the predictions can be made as follows:

```
from sklearn.metrics import roc_auc_score
gb_pred=gb_tree.predict_proba(X_test)
roc_auc_score(y_test, gb_pred[:,1])
```

Implementing GBM in R

GBM in R has similar parameters to GBM in Python. GBM can be implemented as follows:

```
gbm(formula = formula(data),
    distribution = "bernoulli",
    data = list(),
    weights,
    var.monotone = NULL,
    n.trees = 100,
    interaction.depth = 1,
    n.minobsinnode = 10,
    shrinkage = 0.001,
    bag.fraction = 0.5,
    train.fraction = 1.0,
    cv.folds=0,
    keep.data = TRUE,
    verbose = "CV",
    class.stratify.cv=NULL,
    n.cores = NULL)
```

In that formula, we specify the dependent and independent variables in the following fashion: dependent_variable ~ the set of independent variables to be used.

The distribution specifies whether it is a Gaussian, Bernoulli, or AdaBoost algorithm.

- n. trees specifies the number of trees to be built.

- interaction.depth is the max_depth of the trees.

- n.minobsinnode is the minimum number of observations in a node.

- shrinkage is the learning rate.

- bag.fraction is the fraction of the training set observations randomly selected to propose the next tree in the expansion.

GBM algorithm in R is run as follows:

```
library(gbm)
gb=gbm(SeriousDlqin2yrs~.,data=train,n.trees=10,interaction.depth=5,
shrinkage=0.05)
```

The predictions can be made as follows:

```
pred=predict(gb,test,n.trees=10,type="response")
```

Summary

In this chapter, you learnt the following:

- GBM is a decision tree–based algorithm that tries to predict the residual of the previous decision tree in a given decision tree.

- Shrinkage and depth are some of the more important parameters that need to be tuned within GBM.

- The difference between gradient boosting and adaptive boosting.

- How tuning the learning rate parameter improves prediction accuracy in GBM.

CHAPTER 7

Artificial Neural Network

Artificial neural network is a supervised learning algorithm that leverages a mix of multiple hyper-parameters that help in approximating complex relation between input and output. Some of the hyper-parameters in artificial neural network include the following:

- Number of hidden layers

- Number of hidden units

- Activation function

- Learning rate

In this chapter, you will learn the following:

- Working details of neural networks

- The impact of various hyper-parameters on neural networks

- Feed-forward and back propagation

- The impact of learning rate on weight updates

- Ways to avoid Over-fitting in neural networks

- How to implement neural network in Excel, Python, and R

Neural networks came about from the fact that not everything can be approximated by a linear/logistic regression—there may be potentially complex shapes within data that can only be approximated by complex functions. The more complex the function (with some way to take care of overfitting), the better is the accuracy of predictions. We'll start by looking at how neural networks work toward fitting data into a model.

© V Kishore Ayyadevara 2018
V. K. Ayyadevara, *Pro Machine Learning Algorithms*, https://doi.org/10.1007/978-1-4842-3564-5_7

Structure of a Neural Network

The typical structure of a neural network is shown in Figure 7-1.

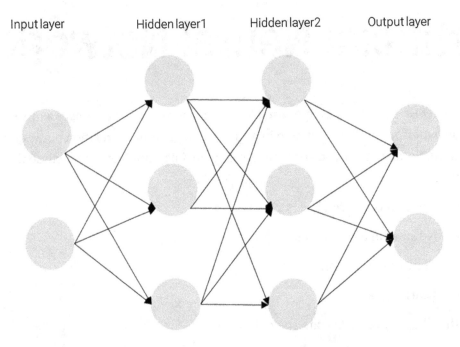

Figure 7-1. *Neural network structure*

The input level/layer in the figure is typically the independent variables that are used to predict the output (dependent variable) level/layer. Typically in a regression problem, there will be only one node in output layer and in a classification problem, the output layer contains as many nodes as the number of classes (distinct values) present in dependent variable. The hidden level/layer is used to transform the input variables into a higher order function. The way the hidden layer transforms the output is shown in Figure 7-2.

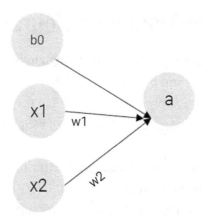

Figure 7-2. *Transforming the output*

In Figure 7.2, *x1* and *x2* are the independent variables, and *b0* is the bias term (similar to the bias in linear/logistic regression). *w1* and *w2* are the weights given to each of the input variables. If *a* is one of the units/neurons in hidden layer, it is equal to the following:

$$a = f\left(\sum_{i=0}^{N} w_i x_i \right)$$

The function in the preceding equation is the *activation* function we are applying on top of the summation so that we attain non-linearity (we need non-linearity so that our model can now learn complex patterns). The different activation functions are discussed in more detail in a later section.

Moreover, having more than one hidden layer helps in achieving high non-linearity. We want to achieve high non-linearity because without it, a neural network would be a giant linear function.

Hidden layers are necessary when the neural network has to make sense of something very complicated, contextual, or non-obvious, like image recognition. The term *deep learning* comes from having many hidden layers. These layers are known as *hidden* because they are not visible as a network output.

Working Details of Training a Neural Network

Training a neural network basically means calibrating all the weights by repeating two key steps: forward propagation and back propagation.

In *forward propagation*, we apply a set of weights to the input data and calculate an output. For the first forward propagation, the set of weights' values are initialized randomly.

In *back propagation*, we measure the margin of error of the output and adjust the weights accordingly to decrease the error.

Neural networks repeat both forward and back propagation until the weights are calibrated to accurately predict an output.

Forward Propagation

Let's go through a simple example of training a neural network to function as an *exclusive or* (*XOR*) operation to illustrate each step in the training process. The XOR function can be represented by the mapping of the inputs and outputs, as shown in the following table, which we'll use as training data. It should provide a correct output given any input acceptable by the XOR function.

Input	Output
(0,0)	0
(0,1)	1
(1,0)	1
(1,1)	0

Let's use the last row from the preceding table, (1,1) => 0, to demonstrate forward propagation, as shown in Figure 7-3. Note that, while this is a classification problem, we will still treat it as a regression problem, only to understand how forward and back propagation work.

Input layer   Hidden layer   Output layer

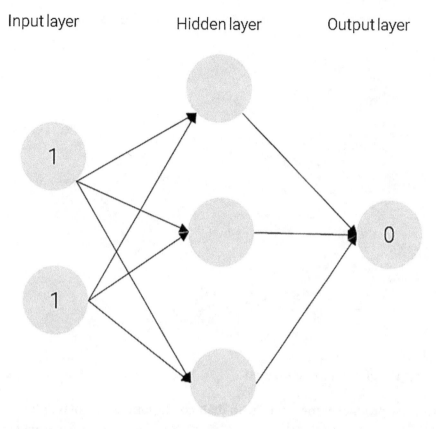

Figure 7-3. *Applying a neural network*

We now assign weights to all the synapses. Note that these weights are selected randomly (the most common way is based on Gaussian distribution) since it is the first time we're forward propagating. The initial weights are randomly assigned to be between 0 and 1 (but note that the final weights don't need to be between 0 and 1), as shown in Figure 7-4.

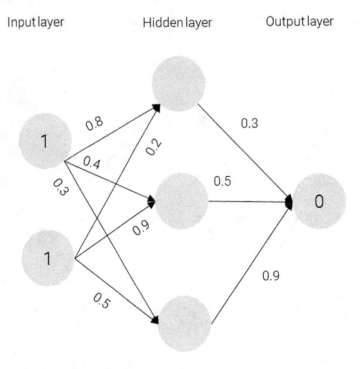

Figure 7-4. *Weights on the synapses*

We sum the product of the inputs with their corresponding set of weights to arrive at the first values for the hidden layer (Figure 7-5). You can think of the weights as measures of influence that the input nodes have on the output:

$$1 \times 0.8 + 1 \times 0.2 = 1$$

$$1 \times 0.4 + 1 \times 0.9 = 1.3$$

$$1 \times 0.3 + 1 \times 0.5 = 0.8$$

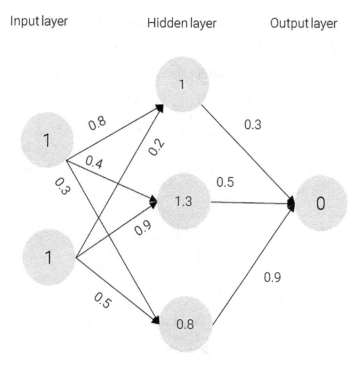

Figure 7-5. *Values for the hidden layer*

Applying the Activation Function

Activation functions are applied at the hidden layer of a neural network. The purpose of an activation function is to transform the input signal into an output signal. They are necessary for neural networks to model complex non-linear patterns that simpler models might miss.

Some of the major activation functions are as follows:

$$\text{Sigmoid} = 1/\left(1 + e^{-x}\right)$$

$$\text{Tanh} = \frac{e^x - e^{-x}}{e^x + e^{-x}}$$

Rectified linear unit = x if $x > 0$, else 0

For our example, let's use the sigmoid function for activation. And applying Sigmoid(x) to the three hidden layer *sums*, we get Figure 7-6:

$$\text{Sigmoid}(1.0) = 0.731$$

$$\text{Sigmoid}(1.3) = 0.785$$

$$\text{Sigmoid}(0.8) = 0.689$$

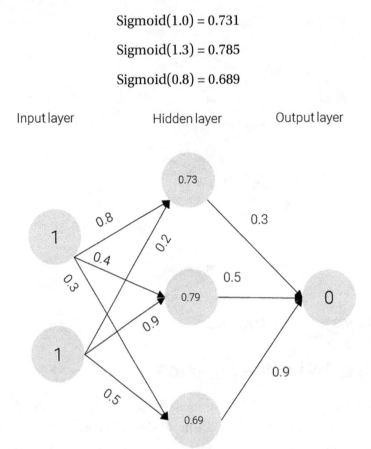

Figure 7-6. *Applying sigmoid to the hidden layer sums*

Then we sum the product of the hidden layer results with the second set of weights (also determined at random the first time around) to determine the output sum:

$$0.73 \times 0.3 + 0.79 \times 0.5 + 0.69 \times 0.9 = 1.235$$

Input layer   Hidden layer   Output layer

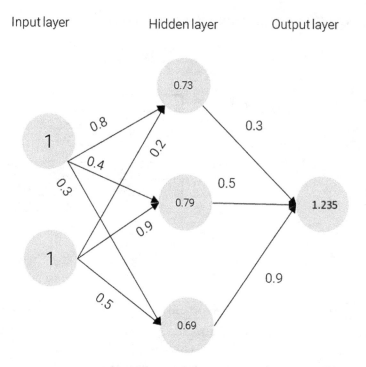

Figure 7-7. *Applying the activation function*

Because we used a random set of initial weights, the value of the output neuron is off the mark—in this case, by 1.235 (since the target is 0).

In Excel, the preceding would look as follows (the excel is available as "NN.xlsx" in github):

1. The input layer has two inputs (1,1), thus input layer is of dimension of 1 × 2 (because every input has two different values).

2. The 1 × 2 hidden layer is multiplied with a randomly initialized matrix of dimension 2 × 3.

3. The output of input to hidden layer is a 1 × 3 matrix:

Input layer
1
1

Pre-activation		
1	1.3	0.8

Hidden layer weights		
0.8	0.4	0.3
0.2	0.9	0.5

Activation		
0.731059	0.785835	0.689974

The formulas for the preceding outputs are as follows:

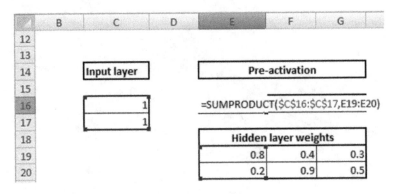

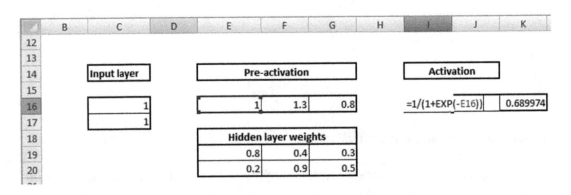

The output of the activation function is multiplied by a 3 × 1 dimensional randomly initialized matrix to get an output that is 1 × 1 in dimension:

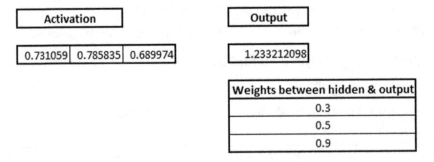

The way to obtain the preceding output is per the following formula:

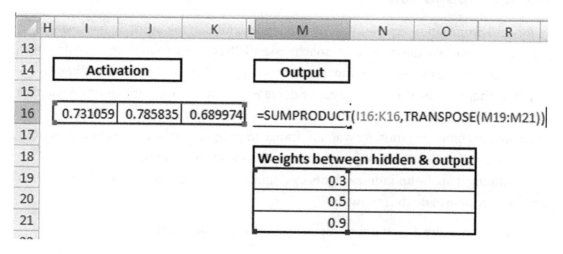

Again, while this is a classification exercise, where we use cross entropy error as loss function, we will still use squared error loss function *only to make the back propagation calculations easier to understand*. We will understand about how classification works in neural network in a later section.

Once we have the output, we calculate the squared error (Overall error) - which is $(1.233-0)^2$, as follows:

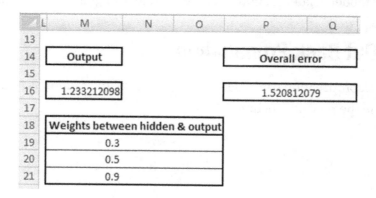

The various steps involved in obtaining squared error from an input layer, collectively form a forward propagation.

Back Propagation

In forward propagation, we took a step from input to hidden to output. In *backward propagation*, we take the reverse approach: essentially, change each weight starting from the last layer by a small amount until the minimum possible error is reached. When a weight is changed, the overall error either decreases or increases. Depending on whether error increased or decreased, the direction in which a weight is updated is decided. Moreover, in some scenarios, for a small change in weight, error increases/decreases by quite a bit, and in some cases error changes only by a small amount.

Summing it up, by updating weights by a small amount and measuring the change in error, we are able to do the following:

1. Decide the direction in which weight needs to be updated

2. Decide the magnitude by which the weights need to be updated

Before proceeding with implementing back propagation in Excel, let's look at one additional aspect of neural networks: the *learning rate*. Learning rate helps us in building trust in our decision of weight updates. For example, while deciding on the magnitude of weight update, we would potentially not change everything in one go but rather take a more careful approach in updating the weights more slowly. This results in obtaining stability in our model. A later section discusses how learning rate helps in stability.

Working Out Back Propagation

To see how back propagation works, let's look at updating the randomly initialized weight values in the previous section.

The overall error of the network with the randomly initialized weight values is 1.52. Let's change the weight between hidden to output layer from 0.3 to 0.29 and see the impact on overall error:

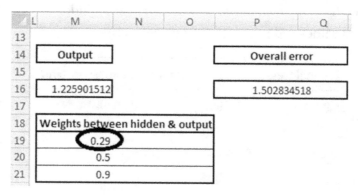

Note that with a small decrease in weight, the overall error decreases from 1.52 to 1.50. Thus, from the two points mentioned earlier, we conclude that 0.3 needs to be reduced to a lower number. The question we need to answer after deciding the direction in which the weight needs to be updated is: "What is the magnitude of weight update?"

If error decreases by a lot from changing the weight by a small amount (0.01), then potentially, weights can be updated by a bigger amount. But if error decreases only by a small amount when weights get updated by a small amount, then weights need to be updated slowly. Thus the weight with a value of 0.3 between the hidden to output layer gets updated as follows:

$$0.3 - 0.05 \times \text{(Reduction in error because of change in weight)}$$

The 0.05 there is the learning parameter, which is to be given as input by user - more on learning rate in the next section. Thus the weight value gets updated to $0.3 - 0.05 \times ((1.52 - 1.50) / 0.01) = 0.21$.

Similarly, the other weights get updated to 0.403 and 0.815:

L	M	N	O	P	Q
13					
14	Output			Overall error	
15					
16	1.032543002			1.066145051	
17					
18	Weights between hidden & output				
19	0.21				
20	0.403				
21	0.815				

Note that the overall error decreased by quite a bit by just changing the weights connecting the hidden layer to the output layer.

Now that we have updated the weights in one layer, we'll update weights that exist in the earlier part of the network—that is, between input and hidden layer activation. Let's change the weight values and calculate the change in error:

Original weight	Updated weight	Decrease in error
0.8	0.7957	0.0009
0.4	0.3930	0.0014
0.3	0.2820	0.0036
0.2	0.1957	0.0009
0.9	0.8930	0.0014
0.5	0.4820	0.0036

Given that error is decreasing every time when the weights are decreased by a small value, we will reduce all the weights to the value calculated above.

Now that the weights are updated, note that the overall error decreased from 1.52 to 1.05. We keep repeating the forward and backward propagation until the overall error is minimized as much as possible.

Stochastic Gradient Descent

Gradient descent is the way in which error is minimized in the scenario just discussed. *Gradient* stands for difference (the difference between actual and predicted) and *descent* means to reduce. *Stochastic* stands for the subset of training data considered to calculate error and thereby weight update (more on the subset of data in a later section).

Diving Deep into Gradient Descent

To further our understanding of gradient descent neural networks, let's start with a known function and see how the weights could be derived: for now, we will have the known function as $y = 6 + 5x$.

The dataset would look as follows (available as "gradient descent batch size.xlsx" in github):

x	y
1	11
2	16
3	21
4	26
5	31
6	36
7	41
8	46
9	51
10	56

Let's randomly initialize the parameters *a* and *b* to values of 2 and 3 (the ideal values of which are 5 and 6). The calculations of weight updates would be as follows:

x	y	a_estimate	b_estimate	y_estimate	squared_error	error_change_a	delta_error_a	new_a	error_change_b	delta_error_b	new_b
1	11	2.00	3.00	5.00	36.00	35.88	11.99	2.12	35.88	11.99	3.12
2	16	2.12	3.12	8.36	58.37	58.22	15.27	2.27	58.07	30.52	3.43
3	21	2.27	3.43	12.55	71.44	71.27	16.89	2.44	70.93	50.62	3.93
4	26	2.44	3.93	18.17	61.36	61.20	15.66	2.60	60.73	62.50	4.56
5	31	2.60	4.56	25.38	31.58	31.47	11.23	2.71	31.02	55.95	5.12
6	36	2.71	5.12	33.41	6.73	6.68	5.18	2.76	6.42	30.77	5.42
7	41	2.76	5.42	40.73	0.07	0.07	0.54	2.77	0.04	3.33	5.46
8	46	2.77	5.46	46.42	0.18	0.19	-0.85	2.76	0.25	-7.40	5.38
9	51	2.76	5.38	51.20	0.04	0.05	-0.42	2.75	0.09	-4.50	5.34
10	56	2.75	5.34	56.13	0.02	0.02	-0.28	2.75	0.05	-3.68	5.30

Note that we started with the random initialization of a_estimate and b_estimate estimate with 2 and 3 (row 1, columns 3 and 4).

We calculate the following:

- Calculate the estimate of *y* using the randomly initialized values of a and b: 5.

- Calculate the squared error corresponding to the values of *a* and *b* (36 in row 1).

- Change the value of *a* slightly (increase it by 0.01) and calculate the squared error corresponding to the changed *a* value. This is stored as the error_change_a column.

- Calculate the change in error in delta_error_a (which is change in error / 0.01). Note that the delta would be very similar if we perform differentiation over loss function with respect to *a*.

- Update the value of *a* based on: new_a = a_estimate + (delta_error_a) × learning_rate.

We consider the learning rate to be 0.01 in this analysis. Do the same analysis for the updated estimate of *b*. Here are the formulas corresponding to the calculations just described:

	K	L	M	N
2				
3	squared error	error_change_a	delta_error_a	new_a
4	=(G4-J4)^2	=(H4+0.01+I4*F4-G4)^2	=(K4-L4)/0.01	=H4+M4*$J$1
5	=(G5-J5)^2	=(H5+0.01+I5*F5-G5)^2	=(K5-L5)/0.01	=H5+M5*$J$1
6	=(G6-J6)^2	=(H6+0.01+I6*F6-G6)^2	=(K6-L6)/0.01	=H6+M6*$J$1
7	=(G7-J7)^2	=(H7+0.01+I7*F7-G7)^2	=(K7-L7)/0.01	=H7+M7*$J$1
8	=(G8-J8)^2	=(H8+0.01+I8*F8-G8)^2	=(K8-L8)/0.01	=H8+M8*$J$1
9	=(G9-J9)^2	=(H9+0.01+I9*F9-G9)^2	=(K9-L9)/0.01	=H9+M9*$J$1
10	=(G10-J10)^2	=(H10+0.01+I10*F10-G10)^2	=(K10-L10)/0.01	=H10+M10*$J$1
11	=(G11-J11)^2	=(H11+0.01+I11*F11-G11)^2	=(K11-L11)/0.01	=H11+M11*$J$1
12	=(G12-J12)^2	=(H12+0.01+I12*F12-G12)^2	=(K12-L12)/0.01	=H12+M12*$J$1
13	=(G13-J13)^2	=(H13+0.01+I13*F13-G13)^2	=(K13-L13)/0.01	=H13+M13*$J$1

	O	P	Q
2			
3	error_change_b	delta_error_b	new_b
4	=(H4+(I4+0.01)*F4-G4)^2	=(K4-O4)/0.01	=I4+P4*$J$1
5	=(H5+(I5+0.01)*F5-G5)^2	=(K5-O5)/0.01	=I5+P5*$J$1
6	=(H6+(I6+0.01)*F6-G6)^2	=(K6-O6)/0.01	=I6+P6*$J$1
7	=(H7+(I7+0.01)*F7-G7)^2	=(K7-O7)/0.01	=I7+P7*$J$1
8	=(H8+(I8+0.01)*F8-G8)^2	=(K8-O8)/0.01	=I8+P8*$J$1
9	=(H9+(I9+0.01)*F9-G9)^2	=(K9-O9)/0.01	=I9+P9*$J$1
10	=(H10+(I10+0.01)*F10-G10)	=(K10-O10)/0.01	=I10+P10*$J$1
11	=(H11+(I11+0.01)*F11-G11)	=(K11-O11)/0.01	=I11+P11*$J$1
12	=(H12+(I12+0.01)*F12-G12)	=(K12-O12)/0.01	=I12+P12*$J$1
13	=(H13+(I13+0.01)*F13-G13)	=(K13-O13)/0.01	=I13+P13*$J$1

Once the values of *a* and *b* are updated (new_a and new_b are calculated in the first row), perform the same analysis on row 2 (Note that we start off row 2 with the updated values of *a* and *b* obtained from the previous row.) We keep on updating the values of *a* and *b* until all the data points are covered. At the end, the updated value of *a* and *b* are 2.75 and 5.3.

Now that we have run through the entire set of data, we'll repeat the whole process with 2.75 and 5.3, as follows:

	x	y	a_estimate	b_estimate	y_estimate	squared_error	error_change_a	delta_error_a	new_a	error_change_b	delta_error_b	new_b
1	11		2.75	5.30	8.05	8.68	8.63	5.88	2.81	8.63	5.88	5.36
2	16		2.81	5.36	13.53	6.10	6.05	4.93	2.86	6.00	9.84	5.46
3	21		2.86	5.46	19.24	3.11	3.08	3.52	2.90	3.01	10.50	5.56
4	26		2.90	5.56	25.15	0.72	0.71	1.69	2.91	0.66	6.65	5.63
5	31		2.91	5.63	31.06	0.00	0.01	(0.13)	2.91	0.01	(0.86)	5.62
6	36		2.91	5.62	36.64	0.41	0.42	(1.29)	2.90	0.49	(8.02)	5.54
7	41		2.90	5.54	41.69	0.47	0.48	(1.38)	2.88	0.57	(10.08)	5.44
8	46		2.88	5.44	46.41	0.16	0.17	(0.82)	2.88	0.24	(7.13)	5.37
9	51		2.88	5.37	51.20	0.04	0.04	(0.40)	2.87	0.08	(4.33)	5.33
10	56		2.87	5.33	56.13	0.02	0.02	(0.26)	2.87	0.05	(3.54)	5.29

The values of *a* and *b* started at 2.75 and 5.3 and ended up at 2.87 and 5.29, which is a little more accurate than the previous iteration. With more iterations, the values of *a* and *b* would converge to the optimal value.

We've looked at the working details of basic gradient descent, but other optimizers perform a similar functionality. Some of them are as follows:

- RMSprop

- Adagrad

- Adadelta

- Adam

- Adamax

- Nadam

Why Have a Learning Rate?

In the scenario just discussed, by having a learning rate of 0.01 we moved the weights from 2,3 to 2.75 and 5.3. Let's look at how the weights would change had the learning rate been 0.05:

x	y	a_estimate	b_estimate	y_estimate	squared_error	error_change_a	delta_error_a	new_a	error_change_b	delta_error_b	new_b
1	11	2.00	3.00	5.00	36.00	35.88	11.99	2.60	35.88	11.99	3.60
2	16	2.60	3.60	9.80	38.46	38.33	12.39	3.22	38.21	24.77	4.84
3	21	3.22	4.84	17.73	10.68	10.61	6.52	3.55	10.48	19.51	5.81
4	26	3.55	5.81	26.80	0.64	0.66	(1.61)	3.46	0.70	(6.56)	5.49
5	31	3.46	5.49	30.89	0.01	0.01	0.20	3.48	0.00	0.81	5.53
6	36	3.48	5.53	36.63	0.40	0.41	(1.28)	3.41	0.48	(7.97)	5.13
7	41	3.41	5.13	39.31	2.86	2.83	3.37	3.58	2.63	23.19	6.29
8	46	3.58	6.29	53.88	62.12	62.28	(15.77)	2.79	63.39	(126.75)	(0.05)
9	51	2.79	(0.05)	2.34	2,367.39	2,366.42	97.30	7.66	2,358.64	874.99	43.70
10	56	7.66	43.70	444.66	151,054.20	151,061.98	(777.32)	(31.21)	151,131.94	(7,774.14)	(345.01)

Note that the moment the learning rate changed from 0.01 to 0.05, in this particular case, the values of *a* and *b* started to have abnormal variations over the latter data points. Thus, a lower learning rate is always preferred. However, note that a lower learning rate would result in a longer time (more iterations) to get the optimal results.

Batch Training

So far, we have seen that the values of *a* and *b* get updated for every row of a dataset. However, that might not be a good idea, as variable values can significantly impact the values of *a* and *b*. Hence, the error calculation is typically done over a batch of data as follows. Let's say the batch size is 2 (in the previous case, batch size was 1):

x	y	a_ estimate	b_ estimate	y_ estimate	squared error	error_ change_a	delta_ error_a	new_a	error_ change_b	delta_ error_b	new_b
1	11	2	3	5	36	35.8801	11.99		35.8801	11.99	
2	16	2	3	8	64	63.8401	15.99		63.6804	31.96	
				Overall	100	99.7202	27.98	2.2798	99.5605	43.95	3.4395

Now for the next batch, the updated values of *a* and *b* are 2.28 and 3.44:

x y	a_ estimate	b_ estimate	y_ estimate	squared error	error_ change_a	delta_ error_a	new_a	error_ change_b	delta_ error_b	new_b
3 21	2.28	3.44	12.60	70.59	70.42	16.79		70.09	50.32	
4 26	2.28	3.44	16.04	99.25	99.05	19.91		98.45	79.54	
			Overall	169.83	169.47	36.71	2.65	168.54	129.86	4.74

The updated values of *a* and *b* are now 2.65 and 4.74, and the iterations continue. Note that, in practice, batch size is at least 32.

The Concept of Softmax

So far, in the Excel implementations, we have performed regression and not classification. They key difference to note when we perform classification is that the output is bound between 0 and 1. In the case of a binary classification, the output layer would have two nodes instead of one. One node corresponds to an output of 0, and the other corresponds to an output of 1.

Now we'll look at how our calculation changes for the discussion in the previous section (where the input is 1,1 and the expected output is 0) when the output layer has two nodes. Given that the output is 0, we will *one-hot-encode* the output as follows: [1,0], where the first index corresponds to an output of 0 and the second index value corresponds to an output of 1.

The weight matrix connecting the hidden layer and the output layer gets changed as follows: instead of a 3 × 1 matrix, it becomes a 3 × 2 matrix, because the hidden layer is now connected to two output nodes (unlike the regression exercise, where it was connected to 1 node):

Activation		
0.731059	0.785835	0.689974

Weights between hidden & output	
0.3	0.1
0.5	0.4
0.9	0.3

Note that because the output nodes are to, our output layer also contains two values, as follows:

Activation		
0.731059	0.785835	0.689974

Output	
1.233212098	0.594432195

Weights between hidden & output	
0.3	0.1
0.5	0.4
0.9	0.3

The one issue with the preceding output is that it has values that are >1 (in other cases, the values could be <0 as well).

Softmax activation comes in handy in such scenario, where the output is beyond the expected value between 0 and 1. Softmax of the above output is calculated as follows:

In Softmax step 1 below, the output is raised to its exponential value. Note that 3.43 is the exponential of 1.233:

Output	
1.233212098	0.594432195

Softmax step 1	
3.43	1.81

In Softmax step 2 below, the softmax output is normalized to get the probabilities in such a way that the sum of probability of both outputs is 1:

Softmax step 1	
3.43	1.81

Softmax step 2	
0.65	0.35

Note that the value of 0.65 is obtained by 3.43 / (3.43 + 1.81).

Now that we have the probability values, instead of calculating the overall squared error, we calculate the cross entropy error, as follows:

1. The final softmax step is compared with actual output:

Softmax step 2	
0.65	0.35

Actual output	
1.00	0

2. The cross entropy error is calculated based on the actual values
    and the predicted values (which are obtained from softmax step 2):

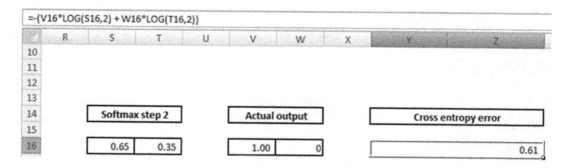

Note the formula for cross entropy error in the formula pane.

Now that we have the final error measure, we deploy gradient descent again to minimize the overall cross entropy error.

Different Loss Optimization Functions

One can optimize for different measures—for example, squared error in regression and cross entropy error in classification. The other loss functions that can be optimized include the following:

- Mean squared error
- Mean absolute percentage error
- Mean squared logarithmic error
- Squared hinge
- Hinge
- Categorical hinge
- Logcosh
- Categorical cross entropy
- Sparse categorical cross entropy
- Binary cross entropy

- Kullback Leibler divergence

- Poisson

- Cosine proximity

Scaling a Dataset

Typically, neural networks perform well when we scale the input datasets. In this section, we'll look at the reason for scaling. To see the impact of scaling on outputs, we'll contrast two scenarios.

Scenario Without Scaling the Input

Scenario	Input	Weight	bias	Sigmoid
1	255	0.01	0	0.93
2	255	0.1	0	1.00
3	255	0.2	0	1.00
4	255	0.3	0	1.00
5	255	0.4	0	1.00
6	255	0.5	0	1.00
7	255	0.6	0	1.00
8	255	0.7	0	1.00
9	255	0.8	0	1.00
10	255	0.9	0	1.00

In the preceding table, various scenarios are calculated where the input is always the same, 255, but the weight that multiplies the input is different in each scenario. Note that the sigmoid output does not change, even though the weight varies by a lot. That's because the weight is multiplied by a large number, the output of which is also a large number.

Scenario with Input Scaling

In this scenario, we'll multiply different weight values by a small input number, as follows:

Scenario	Input	Weight	bias	Sigmoid
1	1	0.01	0	0.50
2	1	0.1	0	0.52
3	1	0.2	0	0.55
4	1	0.3	0	0.57
5	1	0.4	0	0.60
6	1	0.5	0	0.62
7	1	0.6	0	0.65
8	1	0.7	0	0.67
9	1	0.8	0	0.69
10	1	0.9	0	0.71

Now that weights are multiplied by a smaller number, sigmoid output differs by quite a bit for differing weight values.

The problem with a high magnitude of independent variables is significant as the weights need to be adjusted slowly to arrive at the optimal weight value. Given that the weights get adjusted slowly (per the learning rate in gradient descent), it may take considerable time to arrive at the optimal weights when the input is a high magnitude number. Thus, to arrive at an optimal weight value, it is always better to scale the dataset first so that we have our inputs as a small number.

Implementing Neural Network in Python

There are several ways of implementing neural network in Python. Here, we'll look at implementing neural network using the keras framework. You must install tensorflow/theano, and keras before you can implement neural network.

1. Download the dataset and extract the train and test dataset (code available as "NN.ipynb" in github)

```
from keras.datasets import mnist
import matplotlib.pyplot as plt
%matplotlib inline
# load (downloaded if needed) the MNIST dataset
```

```
(X_train, y_train), (X_test, y_test) = mnist.load_data()
# plot 4 images as gray scale
plt.subplot(221)
plt.imshow(X_train[0], cmap=plt.get_cmap('gray'))
plt.subplot(222)
plt.imshow(X_train[1], cmap=plt.get_cmap('gray'))
plt.subplot(223)
plt.imshow(X_train[2], cmap=plt.get_cmap('gray'))
plt.subplot(224)
plt.imshow(X_train[3], cmap=plt.get_cmap('gray'))
# show the plot
plt.show()
```

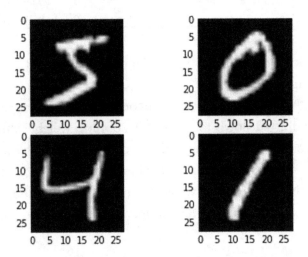

Figure 7-8. *The output*

2. Import the relevant packages:

```
import numpy as np
from keras.datasets import mnist
from keras.models import Sequential
from keras.layers import Dense
from keras.layers import Dropout
from keras.utils import np_utils
```

3. Pre-process the dataset:

```
num_pixels = X_train.shape[1] * X_train.shape[2]
# reshape the inputs so that they can be passed to the
vanilla NN
X_train = X_train.reshape(X_train.shape[0],num_pixels
).astype('float32')
X_test = X_test.reshape(X_test.shape[0],num_pixels).
astype('float32')
# scale inputs
X_train = X_train / 255
X_test = X_test / 255
# one hot encode the output
y_train = np_utils.to_categorical(y_train)
y_test = np_utils.to_categorical(y_test)
num_classes = y_test.shape[1]
```

4. Build a model:

```
# building the model
model = Sequential()
# add 1000 units in the hidden layer
# apply relu activation in hidden layer
model.add(Dense(1000, input_dim=num_
pixels,activation='relu'))
# initialize the output layer
model.add(Dense(num_classes, activation='softmax'))
# compile the model
model.compile(loss='categorical_crossentropy',
optimizer='adam', metrics=['accuracy'])
# extract the summary of model
model.summary()
```

Layer (type)	Output Shape	Param #
dense_39 (Dense)	(None, 1000)	785000
dense_40 (Dense)	(None, 10)	10010

Total params: 795,010
Trainable params: 795,010
Non-trainable params: 0

5. Run the model:

```
model.fit(X_train, y_train, validation_data=(X_test,
y_test), epochs=5, batch_size=1024, verbose=1)
```

```
Train on 60000 samples, validate on 10000 samples
Epoch 1/5
60000/60000 [==============================] - 1s 19us/step - loss: 0.4960 - acc: 0.8650 - val_loss: 0.2336 - val_acc: 0.9326
Epoch 2/5
60000/60000 [==============================] - 1s 11us/step - loss: 0.2003 - acc: 0.9436 - val_loss: 0.1651 - val_acc: 0.9532
Epoch 3/5
60000/60000 [==============================] - 1s 11us/step - loss: 0.1445 - acc: 0.9598 - val_loss: 0.1332 - val_acc: 0.9622
Epoch 4/5
60000/60000 [==============================] - 1s 11us/step - loss: 0.1111 - acc: 0.9692 - val_loss: 0.1083 - val_acc: 0.9687
Epoch 5/5
60000/60000 [==============================] - 1s 11us/step - loss: 0.0870 - acc: 0.9763 - val_loss: 0.0910 - val_acc: 0.9731
```

Note that as the number of epochs increases, accuracy on the test dataset increases as well. Also, in keras we only need to specify the input dimensions in the first layer, and it automatically figures the dimensions for the rest of the layers.

Avoiding Over-fitting using Regularization

Even though we have scaled the dataset, neural networks are likely to overfit on training dataset, as, the loss function (squared error or cross entropy error) ensures that loss is minimized over increasing number of epochs.

However, while training loss keeps on decreasing, it is not necessary that loss on test dataset is also decreasing. With more number of weights (parameters) in a neural network, the chances of over-fitting on training dataset and thus not generalizing on an unseen test dataset is high.

Let us contrast two scenario using the same neural network architecture on MNIST dataset, where in scenario A we consider 5 epochs and thus less chances of over-fitting, while in scenario B, we consider 100 epochs and thus more chances of over-fitting (code available as "Need for regularization in neural network.ipynb" in github).

We should notice that the difference between training and test dataset accuracy is less in the initial few epochs, but as the number of epochs increase, accuracy on training dataset increases, while the test dataset accuracy might not increase after some epochs.

In the case of our run, we see the following accuracy metrics:

Scenario	Training dataset	Test dataset
5 epochs	97.57%	97.27%
100 epochs	100%	98.28%

Once we plot the histogram of weights (for this example, the weights connecting hidden layer to output layer), we will notice that weights have a higher spread (range) in 100 epochs scenario, when compared to weights in 5 epochs scenario, as shown in the below picture:

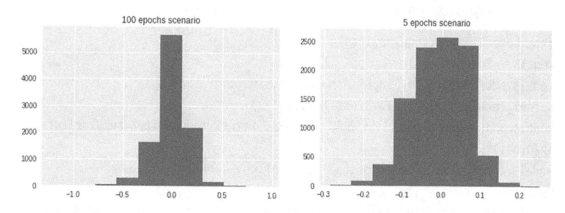

100 epochs scenario had higher weight range, as it was trying to adjust for the edge cases in training dataset over the later epochs, while 5 epochs scenario did not have the opportunity to adjust for the edge cases. While training dataset's edge cases got covered by the weight updates, it is not necessary that test dataset edge cases behave similarly and thus might not have got covered by weight updates. Also, note that an edge case in training dataset can be covered by giving a very high weightage to certain pixels and thus moving quickly towards saturating to either 1 or 0 of the sigmoid curve.

Thus, having high weight values are not desirable for generalization purposes. Regularization comes in handy in such scenario.

Regularization penalizes for having high magnitude of weights. The major types of regularization used are - L1 & L2 regularization

L2 regularization adds the additional cost term to error (loss function) as $\sum w_i^2$

L1 regularization adds the additional cost term to error (loss function) as $\sum |w_i|$

This way, we make sure that weights cannot be adjusted to have a high value so that they work for extreme edge cases in only train dataset.

Assigning Weightage to Regularization term

We notice that our modified loss function, in case of L2 regularization is as follows:

Overall Loss = $\sum (y - \hat{y})^2 + \lambda \sum w_i^2$

where λ is the weightage associated with the regularization term and is a hyper-parameter that needs to be tuned. Similarly, overall loss in case of L1 regularization is as follows:

Overall Loss = $\sum (y - \hat{y})^2 + \lambda \sum |w_i|$

L1/ L2 regularization is implemented in Python as follows:

```
from keras import regularizers
model3 = Sequential()
model3.add(Dense(1000, input_dim=784, activation='relu',
kernel_regularizer=regularizers.l2(0.001)))
model3.add(Dense(10, activation='softmax', kernel_regularizer=regularizers.
l2(0.001)))
model3.compile(loss='categorical_crossentropy', optimizer='adam',
metrics=['accuracy'])
model3.fit(X_train, y_train, validation_data=(X_test, y_test), epochs=100,
batch_size=1024, verbose=2)
```

Notice that, the above involves, invoking an additional hyper-parameter - "kernel_ regularizer" and then specifying whether it is an L1 / L2 regularization. Further we also specify the λ value that gives the weightage to regularization.

We notice that, post regularization, training and test dataset accuracy are similar to each other, where training dataset accuracy is 97.6% while test dataset accuracy is 97.5%. The histogram of weights post L2 regularization is as follows:

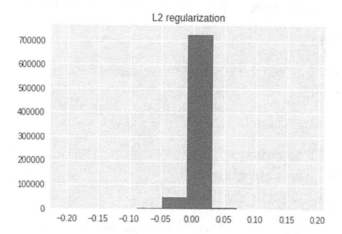

We notice that a majority of weights are now much closer to 0 when compared to the previous two scenarios and thus avoiding the overfitting issue that was caused due to high weight values assigned for edge cases. We would see a similar trend in case of L1 regularization.

Thus, L1 and L2 regularizations help us in avoiding the issue of overfitting on top of training dataset but not generalizing on test dataset.

Implementing Neural Network in R

Similar to the way we implemented a neural network in Python, we will use the keras framework to implement neural network in R. As with Python, multiple packages help us achieve the result.

In order to build neural network models, we will use the kerasR package in R. Given all the dependencies that the kerasR package has on Python, and the need to create a virtual environment, we will perform R implementation in the cloud, as follows (code available as "NN.R" in github):

1. Install the kerasR package:

    ```
    install.packages("kerasR")
    ```

2. Load the installed package:

    ```
    library(kerasR)
    ```

3. Use the MNIST dataset for analysis:

```
mnist <- load_mnist()
```

4. Examine the structure of the mnist object:

```
str(mnist)
```

Note that by default the MNIST dataset has the train and test datasets split.

5. Extract the train and test datasets:

```
mnist <- load_mnist()
X_train <- mnist$X_train
Y_train <- mnist$Y_train
X_test <- mnist$X_test
Y_test <- mnist$Y_test
```

6. Reshape the dataset.

Given that we are performing a normal neural network operation, our input dataset should be of the dimensions (60000,784), whereas X_train is of the dimension (60000,28,28):

```
X_train <- array(X_train, dim = c(dim(X_train)[1], 784))
X_test <- array(X_test, dim = c(dim(X_test)[1], 784))
```

7. Scale the datasets:

```
X_train <- X_train/255
X_test <- X_test/255
```

8. Convert the dependent variables (Y_train and Y_test) into categorical variables:

```
Y_train <- to_categorical(mnist$Y_train, 10)
Y_test <- to_categorical(mnist$Y_test, 10)
```

9. Build a model:

```
model <- Sequential()
model$add(Dense(units = 1000, input_shape = dim(X_train)[2],
activation = "relu"))
model$add(Dense(10,activation = "softmax"))
model$summary()
```

10. Compile the model:

```
keras_compile(model,  loss = 'categorical_crossentropy',
optimizer = Adam(),metrics='categorical_accuracy')
```

11. Fit the model:

```
keras_fit(model, X_train, Y_train,batch_size = 1024,
epochs = 5,verbose = 1, validation_data = list(X_test,
Y_test))
```

The process we just went through should give us a test dataset accuracy of ~98%.

Summary

In this chapter, you learned the following:

- Neural network can approximate complex functions (because of the activation in hidden layers)

- A forward and a backward propagation constitute the building blocks of the functioning of a neural network

- Forward propagation helps us in estimating the error, whereas backward propagation helps in reducing the error

- It is always a better idea to scale the input dataset whenever gradient descent is involved in arriving at the optimal weight values

- L1/ L2 regularization helps in avoiding over-fitting by penalizing for high weight magnitudes

CHAPTER 8

Word2vec

Word2vec is a neural network–based approach that comes in very handy in traditional text mining analysis.

One of the problems with a traditional text mining approach is an issue with the *dimensionality* of data. Given the high number of distinct words within a typical text, the number of columns that are built can become very high (where each column corresponds to a word, and each value in the column specifies whether the word exists in the text corresponding to the row or not—more about this later in the chapter).

Word2vec helps represent data in a better way: words that are similar to each other have similar vectors, whereas words that are not similar to each other have different vectors. In this chapter, we will explore the different ways in which word vectors are calculated.

To get an idea of how Word2vec can be useful, let's explore a problem. Let's say we have two input sentences:

Input sentences

I enjoy playing TT
I like playing TT

Intuitively, we know that *enjoy* and *like* are similar words. However, in traditional text mining, when we *one-hot-encode* the words, our output looks like this:

Unique words

I
enjoy
playing
TT
like

One hot encoding

	I	enjoy	playing	TT	like
I	1	0	0	0	0
enjoy	0	1	0	0	0
playing	0	0	1	0	0
TT	0	0	0	1	0
like	0	0	0	0	1

Notice that one-hot encoding results in each word being assigned a column. The major issue with one-hot-encoding is that the eucledian distance between the words

© V Kishore Ayyadevara 2018
V. K. Ayyadevara, *Pro Machine Learning Algorithms*, https://doi.org/10.1007/978-1-4842-3564-5_8

{*I, enjoy*} is the same as the distance between the words {*enjoy, like*}. But we know that the distance between {*I, enjoy*} should be greater than the distance between {*enjoy, like*} because *enjoy* and *like* are more synonymous to each other.

Hand-Building a Word Vector

Before building a word vector, we'll formulate the hypothesis as follows:

"Words that are related will have similar words surrounding them."

For example, the words *king* and *prince* will have similar words surrounding them more often than not. Essentially, the *context* (the surrounding words) of the words would be similar.

With this hypothesis, let's look at each word as output and all the context words (surrounding words) as input. Thus, our dataset translates as follows (available as "word2vec.xlsx" in github):

	A	B	C	D	E	F	G
1							
2	Input						Output
3							
4	enjoy	playing	TT				I
5	I	playing	TT				enjoy
6	I	enjoy	TT				playing
7	I	enjoy	playing				TT
8	like	playing	TT				I
9	I	playing	TT				like
10	I	like	TT				playing
11	I	like	playing				TT
12							

By using the context words as input, we are trying to predict the given word as output.

A vectorized form of the preceding input and output words looks like this (note that, the column names {I, enjoy, playing, TT, like} are given in row 3 only for reference):

	I	J	K	L	M	N	O	P	Q	R	S	T
1												
2		Vectorizing inputs						Output vector				
3	I	enjoy	playing	TT	like		I	enjoy	playing	TT	like	
4	0	1	1	1	0		1	0	0	0	0	
5	1	0	1	1	0		0	1	0	0	0	
6	1	1	0	1	0		0	0	1	0	0	
7	1	1	1	0	0		0	0	0	1	0	
8	0	0	1	1	1		1	0	0	0	0	
9	1	0	1	1	0		0	0	0	0	1	
10	1	0	0	1	1		0	0	1	0	0	
11	1	0	1	0	1		0	0	0	1	0	

Note that, given the input words—{*enjoy, playing, TT*}—the vector form is {0,1,1,1,0} because the input doesn't contain both *I* and *like,* so the first and last indices are 0 (note the one-hot encoding done in the first page).

For now, let's say we would like to convert the 5-dimensional input vector into a 3-dimensional vector. In such a scenario, our hidden layer has three neurons associated with it. Our neural network would look like Figure 8-1.

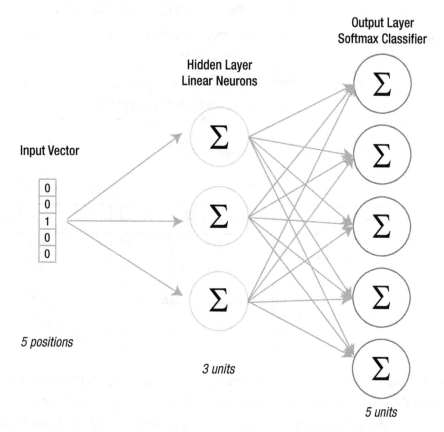

Figure 8-1. *Our neural network*

The dimensions of each layer are as follows:

Layer	Size	Commentary
Input layer	8 × 5	Because there are 8 inputs and 5 indices (unique words)
Weights at hidden layer	5 × 3	Because there are 5 inputs each to the 3 neurons
Output of hidden layer	8 × 3	Matrix multiplication of input and hidden layer
Weights from hidden to output	3 × 5	3 output columns from hidden layer mapped to the 5 original output columns
Output layer	8 × 5	Matrix multiplication between output of hidden layer and the weights from hidden to output layer

The following shows how each of these works out:

A	B	C	D	E	F	G	H	I	J	K	L	M	N	O
1														
2		Vectorizing inputs							Hidden layer				Output of hidden layer	
3														
4	0	1	1	1	0			3.38	-5.78	-0.98		-0.75	6.71	0.23
5	1	0	1	1	0			1.78	3.19	3.63		0.85	-2.26	-4.39
6	1	1	0	1	0			-5.65	-1.66	-0.24		8.27	2.59	-0.52
7	1	1	1	0	0			3.11	5.18	-3.17		-0.48	-4.26	2.41
8	0	0	1	1	1			1.66	3.34	3.76		-0.88	6.87	0.35
9	1	0	1	1	0							0.85	-2.26	-4.39
10	1	0	0	1	1							8.15	2.74	-0.39
11	1	0	1	0	1							-0.61	-4.10	2.54

Note that the input vector is multiplied by a randomly initialized hidden layer weight matrix to obtain the output of hidden layer. Given that the input is 8 × 5 in size and the hidden layer is 5 × 3 in size, the output of matrix multiplication is 8 × 3. And, unlike in a traditional neural network, in the Word2vec approach we don't apply any activations on the hidden layer:

L	M	N	O	P	Q	R	S	T	U	V	W	X	Y	Z	AA
1															
2		Output of hidden layer				weights from hidden to output						Output layer			
3															
4		-0.75	6.71	0.23		-1.95	-2.01	8.71	-1.88	-0.78		53.20	-11.23	5.96	-12.45	-16.48
5		0.85	-2.26	-4.39		7.64	-1.83	1.83	-2.21	-2.49		-27.20	11.52	-1.94	-15.90	11.52
6		8.27	2.59	-0.52		1.88	-2.07	1.19	4.40	-1.49		2.62	-20.28	76.21	-23.55	-12.16
7		-0.48	-4.26	2.41								-27.06	3.76	-9.12	20.96	7.39
8		-0.88	6.87	0.35								54.86	-11.53	5.31	-12.00	-16.96
9		0.85	-2.26	-4.39								-27.20	11.52	-1.94	-15.90	11.52
10		8.15	2.74	-0.39								4.28	-20.58	75.57	-23.10	-12.64
11		-0.61	-4.10	2.54								-25.41	3.46	-9.76	21.41	6.92

Once we have the output of the hidden layer, we multiply them with a matrix of weights from hidden to output layer. Given that the output of hidden layer is 8 × 3 in size and the hidden layer to output is 3 × 5 in size, our output layer is 8 × 5. But note that the output layer has a range of numbers, both positive and negative, as well as numbers that are >1 or <−1.

Hence, just as we did in neural networks, we pass the numbers through a *softmax* to convert them to a number between 0 and 1:

	W	X	Y	Z	AA		AC	AD	AE	AF	AG		AI	AJ	AK	AL	AM
2		Output layer						Softmax part 1						Softmax part 2			
4	53.20	-11.23	5.96	-12.45	-16.48		1.28E+23	1.32E-05	3.86E+02	3.91E-06	6.97E-08		1.00	0.00	0.00	0.00	0.00
5	-27.20	11.52	-1.94	-15.90	11.52		1.53E-12	1.00E+05	1.44E-01	1.25E-07	1.00E+05		0.00	0.50	0.00	0.00	0.50
6	2.62	-20.28	76.21	-23.55	-12.16		1.38E+01	1.56E-09	1.25E+33	5.91E-11	5.21E-06		0.00	0.00	1.00	0.00	0.00
7	-27.06	3.76	-9.12	20.96	7.39		1.77E-12	4.30E+01	1.10E-04	1.26E+09	1.63E+03		0.00	0.00	0.00	1.00	0.00
8	54.86	-11.53	5.31	-12.00	-16.96		6.69E+23	9.83E-06	2.03E+02	6.15E-06	4.33E-08		1.00	0.00	0.00	0.00	0.00
9	-27.20	11.52	-1.94	-15.90	11.52		1.53E-12	1.00E+05	1.44E-01	1.25E-07	1.00E+05		0.00	0.50	0.00	0.00	0.50
10	4.28	-20.58	75.57	-23.10	-12.64		7.21E+01	1.15E-09	6.59E+32	9.29E-11	3.24E-06		0.00	0.00	1.00	0.00	0.00
11	-25.41	3.46	-9.76	21.41	6.92		9.26E-12	3.19E+01	5.78E-05	1.98E+09	1.01E+03		0.00	0.00	0.00	1.00	0.00

For convenience, I have broken down softmax into two steps:

1. Apply exponential to the number.

2. Divide the output of step 1 by the row sum of the output of step 1.

In the preceding output, we see that the output of the first column is very close to 1 in the first row, and the output of the second column is 0.5 in the second row and so on.

Once we obtain the predictions, we compare them with the actuals to calculate the cross entropy loss across the whole batch, as follows:

	AI	AJ	AK	AL	AM	AN	AO	AP	AQ	AR	AS	AT	AU	AV	AW	AX	AY
2		Softmax part 2						Actual output							Cross entropy error		
4	1.00	0.00	0.00	0.00	0.00		1	0	0	0	0		0.00	0.00	0.00	0.00	0.00
5	0.00	0.50	0.00	0.00	0.50		0	1	0	0	0		0.00	-1.00	0.00	0.00	-1.00
6	0.00	0.00	1.00	0.00	0.00		0	0	1	0	0		0.00	0.00	0.00	0.00	0.00
7	0.00	0.00	0.00	1.00	0.00		0	0	0	1	0		0.00	0.00	0.00	0.00	0.00
8	1.00	0.00	0.00	0.00	0.00		1	0	0	0	0		0.00	0.00	0.00	0.00	0.00
9	0.00	0.50	0.00	0.00	0.50		0	0	0	0	1		0.00	-1.00	0.00	0.00	-1.00
10	0.00	0.00	1.00	0.00	0.00		0	0	1	0	0		0.00	0.00	0.00	0.00	0.00
11	0.00	0.00	0.00	1.00	0.00		0	0	0	1	0		0.00	0.00	0.00	0.00	0.00
15															Overall error		4.00

$$\text{Cross entropy loss} = -\sum \text{Actual value} \times \text{Log (probability, 2)}$$

Now that we've calculated the overall cross entropy error, our task is to reduce the overall cross entropy error by varying the weights that are randomly initialized, using an optimizer of choice. Once we arrive at the optimal weight values, we are left with the hidden layer that looks like this:

	D	E	F	G	H	I	J	K	L	M	N	O
4												
5		One hot encoding								Hidden layer		
6		I	enjoy	playing	TT		like					
7	I	1	0	0	0		0			3.38	-5.78	-0.98
8	enjoy	0	1	0	0		0			1.78	3.19	3.63
9	playing	0	0	1	0		0			-5.65	-1.66	-0.24
10	TT	0	0	0	1		0			3.11	5.18	-3.17
11	like	0	0	0	0		1			1.66	3.34	3.76

Now that we have the input words and the hidden layer weights calculated, the words can now be represented in a lower dimension by multiplying the input word with the hidden layer representation.

The matrix multiplication of the input layer (1 × 5 for each word) and hidden layer (5 × 3 weights) is a vector of size (1 × 3):

	D	E	F	G	H	I	J	K	L	M	N	O	P	Q	R	S
4																
5		One hot encoding								Hidden layer				word vector		
6		I	enjoy	playing	TT	like										
7	I	1	0	0	0	0				3.38	-5.78	-0.98		3.38	-5.78	-0.98
8	enjoy	0	1	0	0	0				1.78	3.19	3.63		1.78	3.19	3.63
9	playing	0	0	1	0	0				-5.65	-1.66	-0.24		-5.65	-1.66	-0.24
10	TT	0	0	0	1	0				3.11	5.18	-3.17		3.11	5.18	-3.17
11	like	0	0	0	0	1				1.66	3.34	3.76		1.66	3.34	3.76

If we now consider the words {*enjoy, like*} we should notice that the vectors of the two words are very similar to each other (that is, the distance between the two words is small).

This way, we have converted the original input one-hot-encoded vector, where the distance between {*enjoy, like*} was high to the transformed word vector, where the distance between {*enjoy, like*} is small.

Methods of Building a Word Vector

The method we have adopted in building a word vector in the previous section is called the *continuous bag of words* (*CBOW*) model.

Take a sentence *"The quick brown fox jumped over the dog."* The CBOW model handles that sentence like this:

1. Fix a window size. That is, select *n* words to the left and right of a given word. For example, let's say the window size is 2 words each to the left and right of the given word.

2. Given the window size, the input and output vectors would look like this:

Input words	Output word
{The, quick, fox, jumped}	{brown}
{quick, brown, jumped, over}	{fox}
{brown, fox, over, the}	{jumped}
{fox, jumped, the, dog}	{over}

Another approach to build a word vector is called the *skip-gram* model. In the skip-gram model, the preceding step is reversed, as follows:

Input words	Output word
{brown}	{The, quick, fox, jumped}
{fox}	{quick, brown, jumped, over}
{jumped}	{brown, fox, over, the}
{over}	{fox, jumped, the, dog}

The approach to arrive at the hidden layer vectors remains the same, irrespective of whether it is a skip-gram model or a CBOW model.

Issues to Watch For in a Word2vec Model

For the way of calculation discussed so far, this section looks at some of the common issues we might be facing.

Frequent Words

Typical frequent words like *the* appear quite often in vocabulary. In such cases, the output has the words like *the* occurring a lot more often. If not treated, this might result in a majority of the output being the most frequent words, like *the*, more than other words. We need to have a way to penalize for the number of times a frequently occurring word can be seen in the training dataset.

In a typical Word2vec analysis, the way we penalize for frequently occurring words is as follows. The probability of selecting a word is calculated like this:

$$P(w_i) = \left(\sqrt{\frac{z(w_i)}{0.001}} + 1 \right) \cdot \frac{0.001}{z(w_i)}$$

$z(w)$ is the number of times a word has occurred over the total occurrences of any word. A plot of that formula reveals the curve in Figure 8-2.

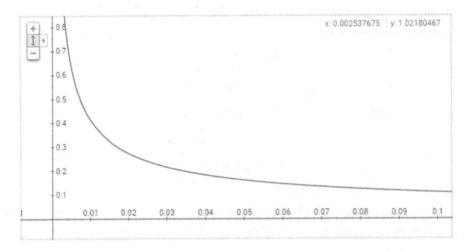

Figure 8-2. *The resultant curve*

Note that as $z(w)$ (x-axis) increases, the probability of selection (y-axis) decreases drastically.

Negative Sampling

Let's assume there are a total of 10,000 unique words in our dataset—that is, each vector is of 10,000 dimensions. Let's also assume that we are creating a 300-dimensional vector from the original 10,000-dimensional vector. This means, from the hidden layer to the output layer, there are a total of $300 \times 10{,}000 = 3{,}000{,}000$ weights.

One of the major issues with such a high number of weights is that it might result in overfitting on top of the data. It also might result in a longer training time.

Negative sampling is one way to overcome this problem. Let's say that instead of checking for all 10,000 dimensions, we pick the index at which the output is 1 (the correct label) and five random indices where the label is 0. This way, we are reducing the number of weights to be updated in a single iteration from 3 million to $300 \times 6 = 1800$ weights.

I said that the selection of negative indices is random, but in actual implementation of Word2vec, the selection is based on the frequency of a word when compared to the other words. The words that are more frequent have a higher chance of getting selected when compared to words that are less frequent.

The probability of selecting five negative words is as follows:

$$P(w_i) = \frac{f(w_i)^{3/4}}{\sum_{j=0}^{n}\left(f(w_j)^{3/4}\right)}$$

$f(w)$ is the frequency of a given word.

Once the probabilities of each word are calculated, the word selection happens as follows: Higher-frequency words are repeated more often, and lower-frequent words are repeated less often and are stored in a table. Given that high-frequency words are repeated more often, the chance of them getting picked up is higher when a random selection of five words is made from the table.

Implementing Word2vec in Python

Word2vec can be implemented in Python using the `gensim` package (the Python implementation is available in github as "word2vec.ipynb").

The first step is to initialize the package:

```
import nltk
import gensim
import pandas as pd
```

Once we import the package, we are expected to provide the parameters discussed in previous sections:

```
import logging
logging.basicConfig(format='%(asctime)s : %(levelname)s : %(message)s',\
    level=logging.INFO)

# Set values for various parameters
num_features = 100     # Word vector dimensionality
min_word_count = 50    # Minimum word count
num_workers = 4        # Number of threads to run in parallel
context = 9            # Context window size
downsampling = 1e-4    # Downsample setting for frequent words
```

- `logging` essentially helps us in tracking the extent to which the word vector calculation is complete.

- `num_features` is the number of neurons in the hidden layer.

- `min_word_count` is the cut-off of the frequency of words that get accepted for the calculation.

- `context` is the window size

- `downsampling` helps in assigning a lower probability of picking the more frequent words.

The input vocabulary for the model should be like the following:

```
sentences[1]
```

```
['maybe',
 'i',
 'just',
 'want',
 'to',
 'get',
 'a',
 'certain']
```

Note that all the input sentences are tokenized.

A Word2vec model is trained as follows:

```
from gensim.models import word2vec
print("Training model...")
w2v_model = word2vec.Word2Vec(t2, workers=num_workers,
          size=num_features, min_count = min_word_count,
          window = context, sample = downsampling)
```

Once a model is trained, the vector of weights for any word in the vocabulary that meets the specified criterion can be obtained as follows:

```
model['word'] # replace the "word" with the word of your interest
```

Similarly, the most similar word to a given word can be obtained like this:

```
model.most_similar('word')
```

Summary

In this chapter, you learned the following:

- Word2vec is an approach that can help convert text words into numeric vectors.

- This acts as a powerful first step for multiple approaches downstream—for example, one can use the word vectors in building a model.

- Word2vec comes up with the vectors using one the CBOW or skip-gram model, which have a neural network architecture that helps in coming up with vectors.

- The hidden layer in neural network is the key to generating the word vectors.

CHAPTER 9

Convolutional Neural Network

In Chapter 7, we looked at a traditional neural network (NN). One of the limitations of a traditional NN is that it is not *translation invariant*—that is, a cat image on the upper right-hand corner of an image would be treated differently from an image that has a cat in the center of the image. *Convolutional neural networks* (CNNs) are used to deal with such issues.

Given that a CNN can deal with translation in images, it is considered a lot more useful and CNN architectures are in fact among the current state-of-the-art techniques in object classification/detection.

In this chapter, you will learn the following:

- Working details of CNN

- How CNN improves over the drawbacks of neural network

- The impact of convolutions and pooling on addressing image translation issues

- How to implement CNN in Python, and R

To better understand the need for CNN further, let's start with an example. Say we would like to classify whether an image has a vertical line in it or not (maybe to tell if the image represents 1 or not). For simplicity's sake, let's assume the image is a 5 × 5 image. Some of the multiple ways in which a vertical line (or a 1) can be written are as follows:

0	0	1	0	0
0	0	1	0	0
0	0	1	0	0
0	0	1	0	0
0	0	1	0	0

0	0	0	1	0
0	0	1	0	0
0	0	1	0	0
0	0	1	0	0
0	1	0	0	0

0	1	0	0	0
0	1	0	0	0
0	0	1	0	0
0	0	1	0	0
0	0	1	0	0

© V Kishore Ayyadevara 2018
V. K. Ayyadevara, *Pro Machine Learning Algorithms*, https://doi.org/10.1007/978-1-4842-3564-5_9

We can also check the different ways in which the digit 1 is written in a MNIST dataset. An image of pixels highlighted for a written 1 is shown in Figure 9-1.

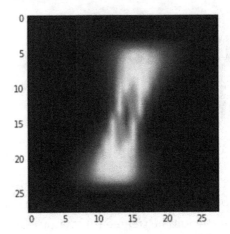

Figure 9-1. *Image of pixels corresponding to images with label 1*

In the image, the redder the pixel, the more often have people written on top of it; bluer means the pixel had been written on fewer times. The pixel in middle is the reddest, quite likely because most people would be writing over that pixel, regardless of the angle they use to write a 1—a vertical line or slanted towards the left or right. In the following section, you would notice that the neural network predictions are not accurate when the image is translated by a few units. In a later section, we will understand how CNN addresses the problem of image translation.

The Problem with Traditional NN

In the scenario just mentioned, a traditional neural network would highlight the image as a *1* only if the pixels around the middle are highlighted and the rest of the pixels in the image are not highlighted (since most people have highlighted the pixels in the middle).

To better understand this problem, let's go through the code we went through in Chapter 7 (code available as "issue with traditional NN.ipynb" in github):

1. Download the dataset and extract the train and test datasets:

```
from keras.datasets import mnist
import matplotlib.pyplot as plt
%matplotlib inline
```

```
# load (downloaded if needed) the MNIST dataset
(X_train, y_train), (X_test, y_test) = mnist.load_data()
# plot 4 images as gray scale
plt.subplot(221)
plt.imshow(X_train[0], cmap=plt.get_cmap('gray'))
plt.subplot(222)
plt.imshow(X_train[1], cmap=plt.get_cmap('gray'))
plt.subplot(223)
plt.imshow(X_train[2], cmap=plt.get_cmap('gray'))
plt.subplot(224)
plt.imshow(X_train[3], cmap=plt.get_cmap('gray'))
# show the plot
plt.show()
```

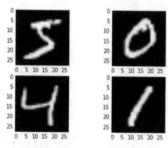

2. Import the relevant packages:

```
import numpy as np
from keras.datasets import mnist
from keras.models import Sequential
from keras.layers import Dense
from keras.layers import Dropout
from keras.layers import Flatten
from keras.layers.convolutional import Conv2D
from keras.layers.convolutional import MaxPooling2D
from keras.utils import np_utils
from keras import backend as K
```

3. Fetch the training set corresponding to the label *1* only:

```
X_train1 = X_train[y_train==1]
```

4. Reshape and normalize the dataset:

```
num_pixels = X_train.shape[1] * X_train.shape[2]
X_train = X_train.reshape(X_train.shape[0],num_pixels
).astype('float32')
X_test = X_test.reshape(X_test.shape[0],num_pixels).
astype('float32')

X_train = X_train / 255
X_test = X_test / 255
```

5. One-hot-encode the labels:

```
y_train = np_utils.to_categorical(y_train)
y_test = np_utils.to_categorical(y_test)
num_classes = y_train.shape[1]
```

6. Build a model and run it:

```
model = Sequential()
model.add(Dense(1000, input_dim=num_pixels, activation='relu'))
model.add(Dense(num_classes, activation='softmax'))
model.compile(loss='categorical_crossentropy', optimizer='adam',
metrics=[''accuracy'])
model.fit(X_train, y_train, validation_data=(X_test, y_test),
epochs=5, batch_size=1024, verbose=1)
```

```
Train on 60000 samples, validate on 10000 samples
Epoch 1/5
60000/60000 [==============================] - 1s 13us/step - loss: 0.4862 - acc: 0.8711 - val_loss: 0.2382 - val_acc: 0.9327
Epoch 2/5
60000/60000 [==============================] - 1s 10us/step - loss: 0.1983 - acc: 0.9446 - val_loss: 0.1686 - val_acc: 0.9503
Epoch 3/5
60000/60000 [==============================] - 1s 11us/step - loss: 0.1423 - acc: 0.9599 - val_loss: 0.1281 - val_acc: 0.9635
Epoch 4/5
60000/60000 [==============================] - 1s 11us/step - loss: 0.1081 - acc: 0.9703 - val_loss: 0.1085 - val_acc: 0.9674
Epoch 5/5
60000/60000 [==============================] - 1s 11us/step - loss: 0.0854 - acc: 0.9769 - val_loss: 0.0957 - val_acc: 0.9719
```

Let's plot what an average *1* label looks like:

```
pic=np.zeros((28,28))
pic2=np.copy(pic)
for i in range(X_train1.shape[0]):
  pic2=X_train1[i,:,:]
```

```
  pic=pic+pic2
pic=(pic/X_train1.shape[0])
plt.imshow(pic)
```

Figure 9-2 shows the result.

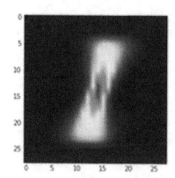

Figure 9-2. *Average 1 image*

Scenario 1

In this scenario, a new image is created (Figure 9-3) in which the original image is translated by 1 pixel toward the left:

```
for i in range(pic.shape[0]):
  if i<20:
    pic[:,i]=pic[:,i+1]
plt.imshow(pic)
```

Figure 9-3. *Average 1 image translated by 1 pixel to the left*

Let's go ahead and predict the label of the image in Figure 9-3 using the built model:

```
model.predict(pic.reshape(1,784))
```

```
array([[0.0000000e+00, 2.8250072e-27, 0.0000000e+00, 0.0000000e+00,
        0.0000000e+00, 0.0000000e+00, 0.0000000e+00, 0.0000000e+00,
        1.0000000e+00, 0.0000000e+00]], dtype=float32)
```

We see the wrong prediction of *8* as output.

Scenario 2

A new image is created (Figure 9-4) in which the pixels are not translated from the original average *1* image:

```
pic=np.zeros((28,28))
pic2=np.copy(pic)
for i in range(X_train1.shape[0]):
  pic2=X_train1[i,:,:]
  pic=pic+pic2
pic=(pic/X_train1.shape[0])
plt.imshow(pic)
```

Figure 9-4. *Average 1 image*

The prediction of this image is as follows:

```
model.predict(pic.reshape(1,784))
```

```
array([[0., 1., 0., 0., 0., 0., 0., 0., 0., 0.]], dtype=float32)
```

We see a correct prediction of *1* as output.

Scenario 3

A new image is created (Figure 9-5) in which the pixels of the original average *1* image are shifted by 1 pixel to the right:

```
pic=np.zeros((28,28))
pic2=np.copy(pic)
for i in range(X_train1.shape[0]):
  pic2=X_train1[i,:,:]
  pic=pic+pic2
pic=(pic/X_train1.shape[0])
pic2=np.copy(pic)
for i in range(pic.shape[0]):
  if ((i>6) and (i<26)):
    pic[:,i]=pic2[:,(i-1)]
plt.imshow(pic)
```

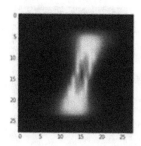

Figure 9-5. *Average 1 image translated by 1 pixel to the right*

Let us go ahead and predict the label of the above image using the built model:

```
model.predict(pic.reshape(1,784))
```

```
array([[0., 1., 0., 0., 0., 0., 0., 0., 0., 0.]], dtype=float32)
```

We have a correct prediction of *1* as output.

Scenario 4

A new image is created (Figure 9-6) in which the pixels of the original average *1* image are shifted by 2 pixels to the right:

```
pic=np.zeros((28,28))
pic2=np.copy(pic)
for i in range(X_train1.shape[0]):
  pic2=X_train1[i,:,:]
  pic=pic+pic2
pic=(pic/X_train1.shape[0])
pic2=np.copy(pic)
for i in range(pic.shape[0]):
  if ((i>6) and (i<26)):
    pic[:,i]=pic2[:,(i-2)]
plt.imshow(pic)
```

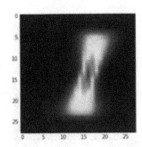

Figure 9-6. *Average 1 image translated by 2 pixels to the right*

We'll predict the label of the image using the built model:

```
model.predict(pic.reshape(1,784))
```

```
array([[0., 0., 0., 1., 0., 0., 0., 0., 0., 0.]], dtype=float32)
```

And we see a wrong prediction of *3* as output.

From the preceding scenarios, you can see that traditional NN fails to produce good results the moment there is translation in the data. These scenarios call for a different way of dealing with the network to address translation variance. And this is where a convolutional neural network (CNN) comes in handy.

Understanding the Convolutional in CNN

You already have a good idea of how a typical neural network works. In this section, let's explore what the word *convolutional* means in CNN. A *convolution* is a multiplication between two matrices, with one matrix being big and the other smaller.

To see convolution, consider the following example.

Matrix A is as follows:

$$\begin{array}{cccc} 1 & 2 & 3 & 4 \\ 5 & 6 & 7 & 8 \\ 9 & 10 & 11 & 12 \\ 13 & 14 & 15 & 16 \end{array}$$

Matrix B is as follows:

$$\begin{array}{cc} 1 & 2 \\ 3 & 4 \end{array}$$

While performing convolution, think of it as sliding the smaller matrix over the bigger matrix: we can potentially come up with nine such multiplications as the smaller matrix is slid over the entire area of the bigger matrix. Note that it is not matrix multiplication:

1. {1,2,5,6} of the bigger matrix is multiplied with {1,2,3,4} of the smaller matrix.

$$1 \times 1 + 2 \times 2 + 5 \times 3 + 6 \times 4 = 44$$

2. {2,3,6,7} of the bigger matrix is multiplied with {1,2,3,4} of the smaller matrix:

$$2 \times 1 + 3 \times 2 + 6 \times 3 + 7 \times 4 = 54$$

3. {3,4,7,8} of the bigger matrix is multiplied with {1,2,3,4} of the smaller matrix:

$$3 \times 1 + 4 \times 2 + 7 \times 3 + 8 \times 4 = 64$$

4. {5,6,9,10} of the bigger matrix is multiplied with {1,2,3,4} of the smaller matrix:

$$5 \times 1 + 6 \times 2 + 9 \times 3 + 10 \times 4 = 84$$

5. {6,7,10,11} of the bigger matrix is multiplied with {1,2,3,4} of the smaller matrix:

$$6 \times 1 + 7 \times 2 + 10 \times 3 + 11 \times 4 = 94$$

6. {7,8,11,12} of the bigger matrix is multiplied with {1,2,3,4} of the smaller matrix:

$$7 \times 1 + 8 \times 2 + 11 \times 3 + 12 \times 4 = 104$$

7. {9,10,13,14} of the bigger matrix is multiplied with {1,2,3,4} of the smaller matrix:

$$9 \times 1 + 10 \times 2 + 13 \times 3 + 14 \times 4 = 124$$

8. {10,11,14,15} of the bigger matrix is multiplied with {1,2,3,4} of the smaller matrix:

$$10 \times 1 + 11 \times 2 + 14 \times 3 + 15 \times 4 = 134$$

9. {11,12,15,16} of the bigger matrix is multiplied with {1,2,3,4} of the smaller matrix:

$$11 \times 1 + 12 \times 2 + 15 \times 3 + 16 \times 4 = 144$$

The result of the preceding steps would be a matrix, as follows:

44	54	64
84	94	104
124	134	144

Conventionally, the smaller matrix is called a *filter* or *kernel*, and the smaller matrix values are arrived at statistically through gradient descent (more on gradient descent a little later). The values within the filter can be considered as the constituent weights.

From Convolution to Activation

In a traditional NN, a hidden layer not only multiplies the input values by the weights, but also applies a non-linearity to the data—it passes the values through an activation function. A similar activity happens in a typical CNN too, where the convolution is passed through an activation function. CNN supports the traditional activations functions we have seen so far: sigmoid, ReLU, Tanh.

For the preceding output, note that the output remains the same when passed through a ReLU activation function, as all the numbers are positive.

From Convolution Activation to Pooling

So far, we have looked at how convolutions work. In this section, we will consider the typical next step after a convolution: pooling.

Let's say the output of the convolution step is as follows (we are not considering the preceding example—this is a new example to illustrate pooling, and the rationale will be explained in a later section):

11	22
32	65

In this case, the output of a convolution step is a 2 × 2 matrix. *Max pooling* considers the 2 × 2 block and gives the maximum value as output—similarly if the output of a convolution step is a bigger matrix, as follows:

11	22	1	2
32	65	3	4
11	12	25	63
13	14	45	32

Max pooling divides the big matrix into non-overlapping blocks of size 2 × 2 each, as follows:

11	22	1	2
32	65	3	4
11	12	25	63
13	14	45	32

From each block, only the element that has the highest value is chosen. So, the output of the max pooling operation on the preceding matrix would be the following:

65	4
14	63

Note that, in practice, it is not necessary to always have a 2 × 2 filter.

The other types of pooling involved are sum and average. Again, in practice we see a lot of max pooling when compared to other types of pooling.

How Do Convolution and Pooling Help?

One of the drawbacks of traditional NN in the MNIST example we looked at earlier was that each pixel is associated with a distinct weight. Thus, if an adjacent pixel other than the original pixel became highlighted, the output would not be very accurate (the example of scenario 1, where the *1*s were slightly to the left of the middle).

This scenario is now addressed, as the pixels share weights that are constituted within each filter. All the pixels are multiplied by all the weights that constitute a filter, and in the pooling layer only the values that are activated the highest are chosen. This way, regardless of whether the highlighted pixel is at the center or is slightly away from the center, the output would more often than not be the expected value. However, the issue remains the same when the highlighted pixels are far away from the center.

Creating CNNs with Code

From the preceding traditional NN scenario, we saw that a NN does not work if the pixels are translated by 1 unit to the left. Practically, we can consider the convolution step as identifying the pattern and pooling step as the one that results in translation variance.

N pooling steps result in at least *N* units of translation invariance. Consider the following example, where we apply one pooling step after convolution (code available as "improvement using CNN.ipynb" in github):

1. Import and reshape the data to fit a CNN:

```
(X_train, y_train), (X_test, y_test) = mnist.load_data()
X_train = X_train.reshape(X_train.shape[0],X_train.shape[1],
X_train.shape[1],1 ).astype('float32')
```

```
X_test = X_test.reshape(X_test.shape[0],X_test.shape[1],X_test.
shape[1],1).astype('float32')

X_train = X_train / 255
X_test = X_test / 255

y_train = np_utils.to_categorical(y_train)
y_test = np_utils.to_categorical(y_test)
num_classes = y_test.shape[1]
Step 2: Build a model
model = Sequential()
model.add(Conv2D(10, (3,3), input_shape=(28, 28,1),
activation='relu'))
model.add(MaxPooling2D(pool_size=(2, 2)))
model.add(Flatten())
model.add(Dense(1000, activation='relu'))
model.add(Dense(num_classes, activation='softmax'))
model.compile(loss='categorical_crossentropy', optimizer='adam',
metrics=['accuracy'])
model.summary()
```

```
Layer (type)                    Output Shape             Param #
=================================================================
conv2d_1 (Conv2D)               (None, 26, 26, 10)       100

max_pooling2d_1 (MaxPooling2     (None, 13, 13, 10)       0

flatten_1 (Flatten)             (None, 1690)             0

dense_1 (Dense)                 (None, 100)              169100

dense_2 (Dense)                 (None, 10)               1010
=================================================================
Total params: 170,210
Trainable params: 170,210
Non-trainable params: 0
```

2. Fit the model:

```
model.fit(X_train, y_train, validation_data=(X_test, y_test),
epochs=5, batch_size=1024, verbose=1)
```

```
Train on 60000 samples, validate on 10000 samples
Epoch 1/5
60000/60000 [==============================] - 3s 50us/step - loss: 0.3802 - acc: 0.8954 - val_loss: 0.1633 - val_acc: 0.9544
Epoch 2/5
60000/60000 [==============================] - 3s 44us/step - loss: 0.1252 - acc: 0.9644 - val_loss: 0.0964 - val_acc: 0.9711
Epoch 3/5
60000/60000 [==============================] - 3s 44us/step - loss: 0.0727 - acc: 0.9795 - val_loss: 0.0691 - val_acc: 0.9780
Epoch 4/5
60000/60000 [==============================] - 3s 44us/step - loss: 0.0483 - acc: 0.9862 - val_loss: 0.0552 - val_acc: 0.9818
Epoch 5/5
60000/60000 [==============================] - 3s 44us/step - loss: 0.0323 - acc: 0.9911 - val_loss: 0.0561 - val_acc: 0.9816
```

For the preceding convolution, where one convolution is followed by one pooling layer, the output prediction works out well if the pixels are translated by 1 unit to the left or right, but does not work when the pixels are translated by more than 1 unit (Figure 9-7):

```
pic=np.zeros((28,28))
pic2=np.copy(pic)
for i in range(X_train1.shape[0]):
  pic2=X_train1[i,:,:]
  pic=pic+pic2
pic=(pic/X_train1.shape[0])
for i in range(pic.shape[0]):
  if i<20:
    pic[:,i]=pic[:,i+1]
plt.imshow(pic)
```

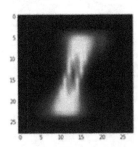

Figure 9-7. *Average 1 image translated by 1 pixel to the left*

Let's go ahead and predict the label of Figure 9-7:

```
model.predict(pic.reshape(1,28,28,1))
```

```
array([[0., 1., 0., 0., 0., 0., 0., 0., 0., 0.]], dtype=float32)
```

We see a correct prediction of *1* as output.

In the next scenario (Figure 9-8), we move the pixels by 2 units to the left:

```
pic=np.zeros((28,28))
pic2=np.copy(pic)
for i in range(X_train1.shape[0]):
  pic2=X_train1[i,:,:]
  pic=pic+pic2
pic=(pic/X_train1.shape[0])
for i in range(pic.shape[0]):
  if i<20:
    pic[:,i]=pic[:,i+2]
plt.imshow(pic)
```

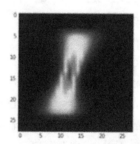

Figure 9-8. *Average 1 image translated by 2 pixels to the left*

Let's predict the label of Figure 9-8 per the CNN model we built earlier:

```
model.predict(pic.reshape(1,28,28,1))
```

```
array([[2.6104576e-16, 0.0000000e+00, 0.0000000e+00, 0.0000000e+00,
        0.0000000e+00, 0.0000000e+00, 1.3525975e-26, 0.0000000e+00,
        1.0000000e+00, 0.0000000e+00]], dtype=float32)
```

We have an incorrect prediction when the image is translated by 2 pixels to the left.

Note that when the number of convolution pooling layers in the model is the same as the amount of translation in an image, the prediction is correct. But prediction is more likely to be incorrect if there are less convolution pooling layers when compared to the translation in image.

Working Details of CNN

Let's build toy CNN code in Python and then implement the outputs in Excel so that it reinforces our understanding (code available as "CNN simple example.ipynb" in github):

1. Import the relevant packages:

```
# import relevant packages
from keras.datasets import mnist
import matplotlib.pyplot as plt
%matplotlib inline
import numpy as np
from keras.datasets import mnist
from keras.models import Sequential
from keras.layers import Dense
from keras.layers import Dropout
from keras.utils import np_utils
from keras.layers import Flatten
from keras.layers.convolutional import Conv2D
from keras.layers.convolutional import MaxPooling2D
from keras.utils import np_utils
from keras import backend as K
from keras import regularizers
```

2. Create a simple dataset:

```
# Create a simple dataset
X_train=np.array([[[1,2,3,4],[2,3,4,5],[5,6,7,8],[1,3,4,5]],
[[-1,2,3,-4],[2,-3,4,5],[-5,6,-7,8],[-1,-3,-4,-5]]])
y_train=np.array([0,1])
```

3. Normalize the inputs by dividing each value with the maximum value in the dataset:

```
X_train = X_train / 8
```

4. One-hot-encode the outputs:

```
y_train = np_utils.to_categorical(y_train)
```

5. Once the simple dataset of just two inputs that are 4 × 4 in size and the two outputs are in place, let's first reshape the input into the required format (which is: number of samples, height of image, width of image, number of channels of the image):

```
X_train = X_train.reshape(X_train.shape[0],X_train.shape[1],
X_train.shape[1],1 ).astype('float32')
```

6. Build a model:

```
model = Sequential()
model.add(Conv2D(1, (3,3), input_shape=(4,4,1),
activation='relu'))
model.add(MaxPooling2D(pool_size=(2, 2)))
model.add(Flatten())
model.add(Dense(10, activation='relu'))
model.add(Dense(2, activation='softmax'))
model.compile(loss='categorical_crossentropy', optimizer='adam',
metrics=['accuracy'])
model.summary()
```

Layer (type)	Output Shape	Param #
conv2d_6 (Conv2D)	(None, 2, 2, 1)	10
max_pooling2d_6 (MaxPooling2	(None, 1, 1, 1)	0
flatten_6 (Flatten)	(None, 1)	0
dense_11 (Dense)	(None, 10)	20
dense_12 (Dense)	(None, 2)	22

```
Total params: 52
Trainable params: 52
Non-trainable params: 0
```

7. Fit the model:

```
model.fit(X_train, y_train, epochs=100, batch_size=2, verbose=1)
```

```
Epoch 95/100
2/2 [==============================] - 0s 2ms/step - loss: 0.2213 - acc: 1.0000
Epoch 96/100
2/2 [==============================] - 0s 2ms/step - loss: 0.2200 - acc: 1.0000
Epoch 97/100
2/2 [==============================] - 0s 2ms/step - loss: 0.2187 - acc: 1.0000
Epoch 98/100
2/2 [==============================] - 0s 2ms/step - loss: 0.2174 - acc: 1.0000
Epoch 99/100
2/2 [==============================] - 0s 2ms/step - loss: 0.2163 - acc: 1.0000
Epoch 100/100
2/2 [==============================] - 0s 3ms/step - loss: 0.2150 - acc: 1.0000
```

The various layers of the preceding model are as follows:

```
model.layers
```

```
[<keras.layers.convolutional.Conv2D at 0x7f97fe6e16d8>,
 <keras.layers.pooling.MaxPooling2D at 0x7f97fe6e1748>,
 <keras.layers.core.Flatten at 0x7f97fe748780>,
 <keras.layers.core.Dense at 0x7f97fe734f98>,
 <keras.layers.core.Dense at 0x7f97fe7340f0>]
```

The name and shape corresponding to various layers are as follows:

```
names = [weight.name for layer in model.layers for weight in layer.weights]
weights = model.get_weights()

for name, weight in zip(names, weights):
    print(name, weight.shape)
```

```
conv2d_6/kernel:0 (3, 3, 1, 1)
conv2d_6/bias:0 (1,)
dense_11/kernel:0 (1, 10)
dense_11/bias:0 (10,)
dense_12/kernel:0 (10, 2)
dense_12/bias:0 (2,)
```

The weights corresponding to a given layer can be extracted as follows:

```
model.layers[0].get_weights()
```

```
[array([[[[ 0.6502779 ]],

         [[ 0.3674555 ]],

         [[-0.04364061]]],


        [[[ 0.8205539 ]],

         [[ 0.5735873 ]],

         [[ 0.13939373]]],


        [[[-0.1292512 ]],

         [[ 0.05793982]],

         [[-0.03353929]]]], dtype=float32),
 array([-0.11086546], dtype=float32)]
```

The prediction for the first input can be calculated as follows:

```
model.predict(X_train[0].reshape(1,4,4,1))
```

```
array([[0.8906642 , 0.10933578]], dtype=float32)
```

Now that we know the probability of 0 for the preceding prediction is 0.89066, let's validate our intuition of CNN so far by matching the preceding prediction in Excel (available as "CNN simple example.xlsx" in github).

The first input and its corresponding scaled version, along with convolution weights and bias (that came out from the model), are as follows:

Input					Scaled input			
1	2	3	4		0.125	0.25	0.375	0.5
2	3	4	5		0.25	0.375	0.5	0.625
5	6	7	8		0.625	0.75	0.875	1
1	3	4	5		0.125	0.375	0.5	0.625

Convolution weights		
0.6503	0.3675	-0.044
0.8206	0.5736	0.1394
-0.129	0.0579	-0.034

Convolution bias	-0.111

The output of convolution is as follows (please check cells L4 to M5 in the 'CNN simple example.xlsx' file):

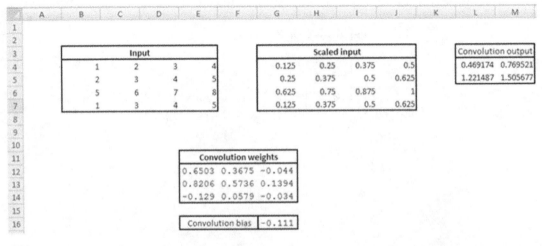

The calculation of convolution is per the following formula:

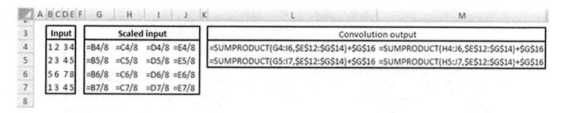

After the convolution layer, we perform the max pooling as follows:

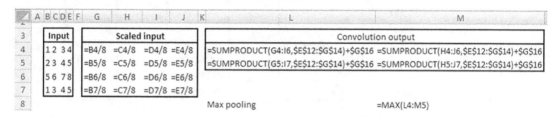

Once the pooling is performed, all the outputs are flattened (per the specification in our model). However, given that our pooling layer has only one output, flattening would also result in a single output.

In the next step, the flattened layer is connected to the hidden dense layer (which in our model specification has ten neurons). The weights and bias corresponding to each neuron are as follows:

Connection from flatten layer to hidden layer	
weight of each neuron	bias of each neuron
-0.36407083	0
-0.33606455	0
0.9516821	-0.09210245
0.95759046	-0.09281213
-0.23468167	0
-0.3994526	0
0.5209155	-0.05047754
-0.36213338	0
-0.27343172	0
-0.05375415	0

The matrix multiplication and the ReLU activation after the multiplication would be as follows:

Max pooling	1.505677818

Connection from flatten layer to hidden layer		Matrix mul	Activation
weight of each neuron	bias of each neuron		
-0.36407083	0	-0.54817337	0
-0.33606455	0	-0.50600494	0
0.9516821	-0.09210245	1.34082418	1.3408242
0.95759046	-0.09281213	1.34901058	1.3490106
-0.23468167	0	-0.35335498	0
-0.3994526	0	-0.60144692	0
0.5209155	-0.05047754	0.73385337	0.7338534
-0.36213338	0	-0.5452562	0
-0.27343172	0	-0.41170008	0
-0.05375415	0	-0.08093643	0

The formulas for the preceding output are as follows:

	K	L	M	N	O	P	Q
8	Max pooling	=MAX(L4:M5)					
9							
10	Connection from flatten layer to hidden layer						
11	weight of each neuron	bias of each neuron		Matrix mul		Activation	
12	-0.36407083	0		=$M$8*L12+M12		=IF(O12>0,O12,0)	
13	-0.33606455	0		=$M$8*L13+M13		=IF(O13>0,O13,0)	
14	0.9516821	-0.09210245		=$M$8*L14+M14		=IF(O14>0,O14,0)	
15	0.95759046	-0.09281213		=$M$8*L15+M15		=IF(O15>0,O15,0)	
16	-0.23468167	0		=$M$8*L16+M16		=IF(O16>0,O16,0)	
17	-0.3994526	0		=$M$8*L17+M17		=IF(O17>0,O17,0)	
18	0.5209155	-0.05047754		=$M$8*L18+M18		=IF(O18>0,O18,0)	
19	-0.36213338	0		=$M$8*L19+M19		=IF(O19>0,O19,0)	
20	-0.27343172	0		=$M$8*L20+M20		=IF(O20>0,O20,0)	
21	-0.05375415	0		=$M$8*L21+M21		=IF(O21>0,O21,0)	

Now let's look at the calculations from hidden layer to output layer. Note that there are two outputs given for each input (output for every row has two columns in dimension: probability of 0 and probability of 1). The weights from hidden layer to output layer are as follows:

Connection from hidden to output	
Weights	
-0.3295091	0.41282815
0.06540197	-0.41082588
0.31798247	-0.6045283
0.04208557	-1.054461
-0.07492512	-0.5271473
-0.61771363	0.27582282
0.3381069	-0.18967173
-0.07473397	-0.0457406
-0.34213108	0.17781854
-0.5068529	0.67832416

bias	-0.502973	0.50297314

Now that each neuron is connected to two weights (where each weight gives its connection to the two outputs), let's look at the calculation from hidden to output layer:

Connection from hidden to output	
Weights	
-0.3295091	0.41282815
0.06540197	-0.41082588
0.31798247	-0.6045283
0.04208557	-1.054461
-0.07492512	-0.5271473
-0.61771363	0.27582282
0.3381069	-0.18967173
-0.07473397	-0.0457406
-0.34213108	0.17781854
-0.5068529	0.67832416

Output layer	
0.22828	-1.86926

bias	-0.502973	0.50297314

The calculation of the output layer is as follows:

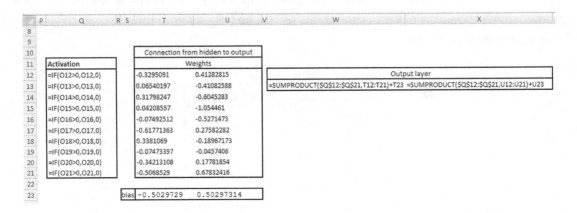

Now that we have some output values, let's calculate the softmax part of the output:

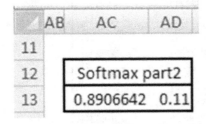

The output would now be exactly the same as what we saw in the output from the keras model:

	AB	AC	AD
11			
12		Softmax part2	
13		0.8906642	0.11

Thus, we have a validation about the intuition laid out in the previous sections.

Deep Diving into Convolutions/Kernels

To see how kernels/filters help, let's go through another scenario. From the MNIST dataset, let's modify the objective in such a way that we are only interested in predicting whether an image is a *1* or not a *1*:

```
(X_train, y_train), (X_test, y_test) = mnist.load_data()
X_train = X_train.reshape(X_train.shape[0],X_train.shape[1],X_train.
shape[1],1 ).astype('float32')
X_test = X_test.reshape(X_test.shape[0],X_test.shape[1],X_test.shape[1],1).
astype('float32')

X_train = X_train / 255
X_test = X_test / 255

X_train1 = X_train[y_train==1]

y_train = np.where(y_train==1,1,0)
y_test = np.where(y_test==1,1,0)
y_train = np_utils.to_categorical(y_train)
y_test = np_utils.to_categorical(y_test)
num_classes = y_test.shape[1]
```

We will come up with a simple CNN where there are only two convolution filters:

```
model = Sequential()
model.add(Conv2D(2, (3,3), input_shape=(28, 28,1), activation='relu'))
model.add(Flatten())
model.add(Dense(1000, activation='relu'))
model.add(Dense(num_classes, activation='softmax'))
model.compile(loss='categorical_crossentropy', optimizer='adam',
metrics=['accuracy'])
model.summary()
```

```
Layer (type)                    Output Shape                Param #
=================================================================
conv2d_9 (Conv2D)               (None, 26, 26, 2)           20
_____
flatten_9 (Flatten)             (None, 1352)                0
_____
dense_27 (Dense)                (None, 1000)                1353000
_____
dense_28 (Dense)                (None, 2)                   2002
=================================================================
Total params: 1,355,022
Trainable params: 1,355,022
Non-trainable params: 0
_____
```

Now we'll go ahead and run the model as follows:

```
model.fit(X_train, y_train, validation_data=(X_test, y_test), epochs=5,
batch_size=1024, verbose=1)
```

```
Train on 60000 samples, validate on 10000 samples
Epoch 1/5
60000/60000 [==============================] - 2s 28us/step - loss: 0.0923 - acc: 0.9605 - val_loss: 0.0169 - val_acc: 0.9956
Epoch 2/5
60000/60000 [==============================] - 1s 21us/step - loss: 0.0183 - acc: 0.9949 - val_loss: 0.0109 - val_acc: 0.9967
Epoch 3/5
60000/60000 [==============================] - 1s 21us/step - loss: 0.0126 - acc: 0.9962 - val_loss: 0.0100 - val_acc: 0.9967
Epoch 4/5
60000/60000 [==============================] - 1s 21us/step - loss: 0.0095 - acc: 0.9973 - val_loss: 0.0085 - val_acc: 0.9972
Epoch 5/5
60000/60000 [==============================] - 1s 21us/step - loss: 0.0074 - acc: 0.9977 - val_loss: 0.0080 - val_acc: 0.9976
```

We can extract the weights corresponding to the filters in the following way:

```
model.layers[0].get_weights()
```

Let's manually convolve and apply the activation by using the weights derived in the preceding step (Figure 9-9):

```
from scipy import signal
from scipy import misc
import numpy as np
import pylab
for j in range(2):
    gradd=np.zeros((30,30))
    for i in range(6000):
```

```
    grad = signal.convolve2d(X_train1[i,:,:,0], model.layers[0].get_
    weights()[0].T[j][0])+model.layers[0].get_weights()[1][j]
    grad = np.where(grad<0,0,grad)
    gradd=grad+gradd
grad2=np.where(gradd<0,0,gradd)
pylab.imshow(grad2/6000)
pylab.gray()
pylab.show()
```

In the figure, note that the filter on the left activates a *1* image a lot more than the filter on the right. Essentially, the first filter helps in predicting label *1* more, and the second filter augments in predicting the rest.

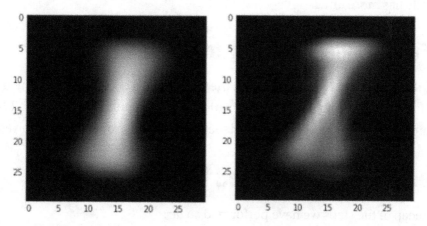

Figure 9-9. *Average filter activations when 1 label images are passed*

From Convolution and Pooling to Flattening: Fully Connected Layer

The outputs we have seen so far until pooling layer are images. In traditional neural network, we would consider each pixel as an independent variable. This is precisely what we are going to perform in the flattening process.

Each pixel of the image is unrolled, and so the process is called *flattening*. For example, the output image after convolution and pooling looks like this:

$$\begin{array}{ccc} 1 & 2 & 3 \\ 0 & 4 & 5 \\ 7 & 4 & 3 \end{array}$$

The output of flattening looks like this:

$$\begin{array}{ccccccccc} 1 & 2 & 3 & 0 & 4 & 5 & 7 & 4 & 3 \end{array}$$

From One Fully Connected Layer to Another

In a typical neural network, the input layer is connected to the hidden layer. In a similar manner, in a CNN the fully connected layer is connected to another fully connected layer that typically has more units.

From Fully Connected Layer to Output Layer

Similar to the traditional NN architecture, the hidden layer is connected to the output layer and is passed through a sigmoid activation to get the output as a probability. An appropriate loss function is also chosen, depending on the problem being solved.

Connecting the Dots: Feed Forward Network

Here is a recap of the steps we have performed so far:

1. Convolution

2. Pooling

3. Flattening

4. Hidden layer

5. Calculating output probability

A typical CNN looks is shown in Figure 9-10 (the most famous—the one developed by the inventor himself, LeNet, as an example):

The subsample written in Figure 9-10 is equivalent to the max pooling step we saw earlier.

Other Details of CNN

In Figure 9-10, we see that the conv1 step has six *channels* or convolutions of the original image. Let's look at this in detail:

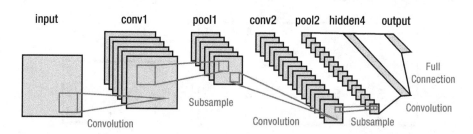

Figure 9-10. *A LeNet*

1. Let's say we have a greyscale image that is 28 × 28 in dimension. Six filters that are 3 × 3 in size would generate images that are 26 × 26 in size. Thus, we are left with six images of size 26 × 26.

2. A typical color image would have three channels (RGB). For simplicity, we can assume that the output image we had in step 1 has six channels - one each for the six filters (though we can't name them as RGB like the three-channel version). In this step, we would perform max pooling on each of the six channels separately. This would result in six images (channels) that are 13 × 13 in dimension.

3. In the next convolution step, we multiply the six channels of 13 × 13 images with weights of dimensions 3 × 3 × 6. That's a 3-dimensional weight matrix convolving over a 3-dimensional image (where the image has dimensions 13 × 13 × 6). This would result in an image of 11 × 11 in dimension for each filter.

   Let's say we've considered ten different weight matrices (cubes, to be precise). This would result in an image that is 11 × 11 × 10 in dimension.

4. Max pooling on each of the 11 × 11 images (which are ten in number) would result in a 5 × 5 image. Note that, when the max pooling is

performed on an image that has odd number of dimensions, pooling gives us the rounded-down image—that is, 11/2 is rounded down to 5.

A *stride* is the amount by which the filter that convolves over the original image moves from one step to the next step. For example, if the stride value is 2, the distance between 2 consecutive convolutions is 2 pixels. When the stride value is 2, the multiplication would happen as follows, where A is the bigger matrix and B is the filter:

1	2	3	4	5
2	3	4	5	6
4	5	6	7	8
5	6	7	8	9
6	7	8	9	10

1	2	3
0	4	5
7	4	3

The first convolution would be between:

1	2	3
2	3	4
4	5	6

1	2	3
0	4	5
7	4	3

The second convolution would be between:

3	4	5
4	5	6
6	7	8

1	2	3
0	4	5
7	4	3

The third convolution would be between:

4	5	6
5	6	7
6	7	8

1	2	3
0	4	5
7	4	3

The final convolution would be between:

6	7	8	1	2	3
7	8	9	0	4	5
8	9	10	7	4	3

Note that the output of the convolution is a 2×2 matrix when the stride is 2 for the matrices of the given dimensions here.

PADDING

Note that the size of the resulting image is reduced when a convolution is performed on top of it. One way to get rid of the size-reduction issue is to *pad* the original image with zeroes on the four borders. This way, a 28×28 image would be translated into a 30×30 image. Thus, when the 30×30 image is convolved by a 3×3 filter, the resulting image would be a 28×28 image.

Backward Propagation in CNN

Backward propagation in CNN is done in similarly to a typical NN, where the impact of changing a weight by a small amount on the overall weight is calculated. But in place of weights, as in NN, we have filters/matrices of weights that need to be updated to minimize the overall loss.

Sometimes, given that there are typically millions of parameters in a CNN, having regularization can be helpful. Regularization in CNN can be achieved using the dropout method or the L1 and L2 regularizations. *Dropout* is done by choosing not to update some weights (typically a randomly chosen 20% of total weights) and training the entire network over the whole number of epochs.

Putting It All Together

The following code implements a three-convolution pooling layer followed by flattening and a fully connected layer:

```
 (X_train, y_train), (X_test, y_test) = mnist.load_data()
X_train = X_train.reshape(X_train.shape[0],X_train.shape[1],X_train.
shape[1],1 ).astype('float32')
X_test = X_test.reshape(X_test.shape[0],X_test.shape[1],X_test.shape[1],1).
astype('float32')

X_train = X_train / 255
X_test = X_test / 255

y_train = np_utils.to_categorical(y_train)
y_test = np_utils.to_categorical(y_test)
num_classes = y_test.shape[1]
```

In the next step, we build the model, as follows:

```
model = Sequential()
model.add(Conv2D(32, (3,3), input_shape=(28, 28,1), activation='relu'))
model.add(MaxPooling2D(pool_size=(2, 2)))
model.add(Conv2D(64, (3,3), activation='relu'))
model.add(MaxPooling2D(pool_size=(2, 2)))
model.add(Flatten())
model.add(Dense(1000, activation='relu'))
model.add(Dense(num_classes, activation='softmax'))
model.compile(loss='categorical_crossentropy', optimizer='adam',
metrics=['accuracy'])
model.summary()
```

Layer (type)	Output Shape	Param #
conv2d_16 (Conv2D)	(None, 26, 26, 32)	320
max_pooling2d_7 (MaxPooling2	(None, 13, 13, 32)	0
conv2d_17 (Conv2D)	(None, 11, 11, 64)	18496
max_pooling2d_8 (MaxPooling2	(None, 5, 5, 64)	0
flatten_13 (Flatten)	(None, 1600)	0
dense_35 (Dense)	(None, 1000)	1601000
dense_36 (Dense)	(None, 10)	10010

```
Total params: 1,629,826
Trainable params: 1,629,826
Non-trainable params: 0
```

Finally, we fit the model, as follows:

```
model.fit(X_train, y_train, validation_data=(X_test, y_test), epochs=5,
batch_size=1024, verbose=1)
```

Note that the accuracy of the model trained using the preceding code is ~98.8%. But note that although this model works best on the test dataset, an image that is translated or rotated from the test MNIST dataset would not be classified correctly (In general, CNN could only help when the image is translated by the number of convolution pooling layers). That can be verified by looking at the prediction when the average *1* image is translated by 2 pixels to the left once, and in another scenario, 3 pixels to the left, as follows:

```
pic=np.zeros((28,28))
pic2=np.copy(pic)
for i in range(X_train1.shape[0]):
  pic2=X_train1[i,:,:,0]
  pic=pic+pic2
pic=(pic/X_train1.shape[0])
for i in range(pic.shape[0]):
  if i<20:
    pic[:,i]=pic[:,i+2]
model.predict(pic.reshape(1,28,28,1))
```

```
        array([[2.4316866e-02, 9.0267426e-01, 7.1549327e-03, 2.8638367e-05,
                2.3012757e-03, 3.0512114e-03, 4.0094595e-02, 7.7957986e-04,
                1.8754713e-02, 8.4394397e-04]], dtype=float32)
```

Note that, in this case, where the image is translated by 2 units to the left, the predictions are accurate:

```
pic=np.zeros((28,28))
pic2=np.copy(pic)
for i in range(X_train1.shape[0]):
  pic2=X_train1[i,:,:,0]
  pic=pic+pic2
pic=(pic/X_train1.shape[0])
for i in range(pic.shape[0]):
  if i<20:
    pic[:,i]=pic[:,i+3]
model.predict(pic.reshape(1,28,28,1))
```

```
        array([[4.3927294e-01, 2.5317281e-01, 1.6524114e-02, 9.6656995e-06,
                1.1181158e-02, 9.0220626e-03, 2.3373538e-01, 1.5526810e-03,
                3.4207623e-02, 1.3214557e-03]], dtype=float32)
```

Note that here, when the image is translated by more pixels than convolution pooling layers, the prediction is not accurate. This issue is solved by using *data augmentation,* the topic of the next section.

Data Augmentation

Technically, a translated image is the same as a new image that is generated from the original image. New data can be generated by using the ImageDataGenerator function in keras:

```
from keras.preprocessing.image import ImageDataGenerator
shift=0.2
datagen = ImageDataGenerator(width_shift_range=shift)
datagen.fit(X_train)
i=0
for X_batch,y_batch in datagen.flow(X_train,y_train,batch_size=100):
```

```
  i=i+1
  print(i)
  if(i>500):
    break
  X_train=np.append(X_train,X_batch,axis=0)
  y_train=np.append(y_train,y_batch,axis=0)
print(X_train.shape)
```

From that code, we have generated 50,000 random shufflings from our original data, where the pixels are shuffled by 20%.

As we plot the image of *1* now (Figure 9-11), note that there is a wider spread for the image:

```
y_train1=np.argmax(y_train,axis=1)
X_train1=X_train[y_train1==1]
pic=np.zeros((28,28))
pic2=np.copy(pic)
for i in range(X_train1.shape[0]):
  pic2=X_train1[i,:,:,0]
  pic=pic+pic2
pic=(pic/X_train1.shape[0])
plt.imshow(pic)
```

Figure 9-11. *Average 1 post data augmentation*

Now the predictions will work even when we don't do convolution pooling for the few pixels that are to the left or right of center. However, for the pixels that are far away from the center, correct predictions will come once the model is built using the convolution and pooling layers.

So, data augmentation helps in further generalizing for variations of the image across the image boundaries when using the CNN model, even with fewer convolution pooling layers.

Implementing CNN in R

To implement CNN in R, we will leverage the same package we used to implement neural network in R—kerasR (code available as "kerasr_cnn_code.r" in github):

```
# Load, split, transform and scale the MNIST dataset
mnist <- load_mnist()

X_train <- array(mnist$X_train, dim = c(dim(mnist$X_train), 1)) / 255
Y_train <- to_categorical(mnist$Y_train, 10)
X_test <- array(mnist$X_test, dim = c(dim(mnist$X_test), 1)) / 255
Y_test <- to_categorical(mnist$Y_test, 10)

# Build the model
model <- Sequential()
model$add(Conv2D(filters = 32, kernel_size = c(3, 3),input_shape = c(28, 28, 1)))
model$add(Activation("relu"))
model$add(MaxPooling2D(pool_size=c(2, 2)))
model$add(Flatten())
model$add(Dense(128))
model$add(Activation("relu"))
model$add(Dense(10))
model$add(Activation("softmax"))
# Compile and fit the model
```

```
keras_compile(model,  loss = 'categorical_crossentropy', optimizer =
Adam(),metrics='categorical_accuracy')
keras_fit(model, X_train, Y_train, batch_size = 1024, epochs = 5, verbose = 1,
validation_data = list(X_test,Y_test))
```

The preceding code results in an accuracy of ~97%.

Summary

In this chapter, we saw how convolutions help us identify the structure of interest and how pooling helps ensure that the image is recognized even when translation happens in the original image. Given that CNN is able to adapt to image translation through convolution and pooling, it's in a position to give better results than the traditional neural network.

CHAPTER 10

Recurrent Neural Network

In Chapter 9, we looked at how convolutional neural networks (CNNs) improve upon the traditional neural network architecture for image classification. Although CNNs perform very well for image classification in which image translation and rotation are taken care of, they do not necessarily help in identifying temporal patterns. Essentially, one can think of CNNs as identifying static patterns.

Recurrent neural networks (RNNs) are designed to solve the problem of identifying temporal patterns.

In this chapter, you will learn the following:

- Working details of RNN

- Using embeddings in RNN

- Generating text using RNN

- Doing sentiment classification using RNN

- Moving from RNN to LSTM

RNN can be architected in multiple ways. Some of the possible ways are shown in Figure 10-1.

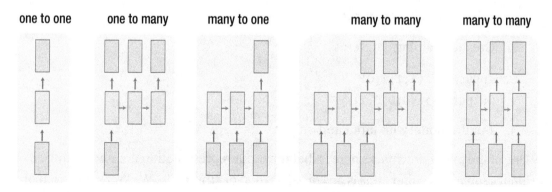

Figure 10-1. *RNN examples*

© V Kishore Ayyadevara 2018
V. K. Ayyadevara, *Pro Machine Learning Algorithms*, https://doi.org/10.1007/978-1-4842-3564-5_10

In Figure 10-1 note the following:

- The boxes in the bottom are inputs

- The boxes in the middle are hidden layers

- The boxes at the top are outputs

An example of the one-to-one architecture shown is a typical neural network that we have looked at in chapter 7, with a hidden layer between the input and the output layer. An example of one-to-many RNN architecture would be to input an image and output the caption of the image. An example of many-to-one RNN architecture might be a movie review given as input and the movie sentiment (positive, negative or neutral review) as output. Finally, an example of many-to-many RNN architecture would be machine translation from one language to another language.

Understanding the Architecture

Let's go through an example and look more closely at RNN architecture. Our task is as follows: "Given a string of words, predict the next word." We'll try to predict the word that comes after "This is an ____". Let's say the actual sentence is "This is an example."

Traditional text mining techniques would solve the problem in the following way:

1. Encode each word, leaving space for an extra word, if needed:

   ```
   This: {1,0,0,0}
   is: {0,1,0,0}
   an: {0,0,1,0}
   ```

2. Encode the sentence:

   ```
   "This is an": {1,1,1,0}
   ```

3. Create a training dataset:

   ```
   Input --> {1,1,1,0}
   Output --> {0,0,0,1}
   ```

4. Build a model with input and output.

One of the major drawbacks here is that the input representation does not change if the input sentence is either "this is an" or "an is this" or "this an is". We know that each of these is very different and cannot be represented by the same structure mathematically.

This realization calls for having a different architecture, one that looks more like Figure 10-2.

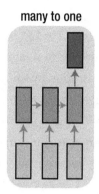

many to one

Figure 10-2. *A change in the architecture*

In the architecture shown in Figure 10-2, each of the individual words in the sentence goes into a separate box among the three input boxes. Moreover, the structure of the sentence is preserved since "this" gets into the first box, "is" gets into the second box, and "an" gets into the third box.

The output "example" is expected in the output box at the top.

Interpreting an RNN

We can think of RNN as a mechanism to hold memory, where the memory is contained within the hidden layer. This is illustrated in Figure 10-3.

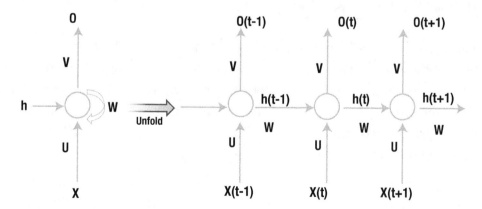

Figure 10-3. *Memory in the hidden layer*

The network on the right in Figure 10-3 is an unrolled version of the network on the left. The network on the left is a traditional one, with one change: the hidden layer is connected to itself along with being connected to the input (the hidden layer is the circle in the figure).

Note that when a hidden layer is connected to itself along with input layer, it is connected to a "previous version" of the hidden layer and the current input layer. We can consider this phenomenon of the hidden layer being connect back to itself as the mechanism by which memory is created in RNN.

The weight U represents the weights that connect the input layer to the hidden layer, the weight W represents the hidden-layer-to-hidden-layer connection, and the weight V represents the hidden-layer-to-output-layer connection.

WHY STORE MEMORY?

There is a need to store memory because, in the preceding example and in text generation in general, the next word does not necessarily rely on the preceding word but the context of the few words preceding the word to predict.

Given that we are looking at the preceding words, there should be a way to keep them in memory so that we can predict the next word more accurately. Moreover, we should also have the memory in order—more often than not, more recent words are more useful in predicting the next word than the words that are far away from the word being predicted.

Working Details of RNN

Note that a typical NN has an input layer followed by an activation in the hidden layer and then a softmax activation at the output layer. RNN is similar, but with memory. Let's look at another example: "This is an example". Given an input "This", we are expected to predict "is" and similarly for an input "is", we are expected to come up with a prediction of "an" and a prediction of "example" for "an" as input. The dataset is available as "RNN dimension intuition.xlsx" in github.

The encoded input and output words are as follows:

C	D	E	F	G	H	I	J	K	L	M	N	O	P
1			This	is	an	example					Expected output		
2		This	1	0	0	0			is	0	1	0	0
3	Input	is	0	1	0	0			an	0	0	1	0
4		an	0	0	1	0			example	0	0	0	1
5		example	0	0	0	1							

The RNN structure looks like Figure 10-4.

many to one

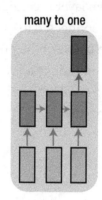

Figure 10-4. *The RNN structure*

Let's deconstruct the dimensions of each weight matrix associated:

C	D	E	F	G	H	I	J	K	L	M	N	O	P
1			This	is	an	example					Expected output		
2		This	1	0	0	0			is	0	1	0	0
3	Input	is	0	1	0	0			an	0	0	1	0
4		an	0	0	1	0			example	0	0	0	1
5		example	0	0	0	1			blank	0	0	0	0
6													
7													
8			0.033	0.021	0.065								
9	wxh		0.052	-0.08	-0.04								
10			0.048	0.04	-0.08								
11			0.056	-0.03	-0.06								
12													
13													
14	Hidden layer		0.033	0.021	0.065								
15			0.052	-0.08	-0.04								
16			0.048	0.04	-0.08								
17			0.056	-0.03	-0.06								

wxh is randomly initialized and 4 × 3 in dimension. Each input is 1 × 4 in dimension. Thus, the hidden layer, which is a matrix multiplication between the input and *wxh*, is 1 × 3 in dimension for each input row. The expected output is the one-hot encoded version of the word that comes next to the input word in our sentence. Note that, the last prediction "blank" is inaccurate because we have all 0s as expected output. Ideally, we would have a new column in the one-hot encoded version that takes care of all the unseen words. However, for the sake of understanding the working details of RNN we will keep it simple with 4 columns in expected output.

As we saw earlier, in RNN, a hidden layer is connected to another hidden layer when unrolled. Given that a hidden layer is connected to the next hidden layer, the weight (*whh*) associated with the connection between the previous hidden layer and the current hidden layer would be 3 × 3 in dimension, since a 1 × 3 matrix multiplied with 3 × 3 matrix would yield a 1 × 3 matrix. Final hidden layer calculations in the below picture are explained in subsequent pages.

wxh					whh			
0.033	0.021	0.065				-0.03	0.043	0.032
0.052	-0.08	-0.04				-0.05	-0.048	0.024
0.048	0.04	-0.08				-0.08	-0.032	0.047
0.056	-0.03	-0.06						

Hidden layer				Final hidden layer			
0.033	0.021	0.065			0.03	0.02	0.07
0.052	-0.08	-0.04			0.05	(0.08)	(0.04)
0.048	0.04	-0.08			0.05	0.05	(0.08)
0.056	-0.03	-0.06			0.06	(0.03)	(0.06)

Note that, *wxh* and *whh* are random initializations, whereas the hidden layer and the final hidden layer are calculated. We will look at how the calculations are done in the following pages.

The calculation for the hidden layer at various time steps is performed as follows:

$$h^{(t)} = \phi_h\left(z_h^{(t)}\right) = \phi_h\left(W_{xh}x^{(t)} + W_{hh}h^{(t-1)}\right)$$

where ϕ_h is an activation that is performed (tanh activation in general). Calculation from the input layer to the hidden layer consists of two components:

- Matrix multiplication of the input layer and *wxh*.

- Matrix multiplication of hidden layer and *whh*.

Final calculation of the hidden layer value at a given time step would be the summation of the preceding two matrix multiplications and passing the result through a tanh activation function.

Matrix multiplication of the input layer and *wxh* is shown here:

C	D	E	F	G	H	I	J	K
1			This	is	an	example		
2		This	1	0	0	0		
3	Input	is	0	1	0	0		
4		an	0	0	1	0		
5		example	0	0	0	1		
6								
7								
8			0.033	0.021	0.065			
9	wxh		0.052	-0.08	-0.04			
10			0.048	0.04	-0.08			
11			0.056	-0.03	-0.06			
12								
13								
14	Hidden layer		0.033	0.021	=SUMPRODUCT($F2:$I2,TRANSPOSE(H$8:H$11))			
15			0.052	-0.08	-0.04			
16			0.048	0.04	-0.08			
17			0.056	-0.03	-0.06			

The following sections go through the calculation of the hidden layer value at different time steps.

Time Step 1

The hidden layer value at the first time step would be the value of matrix multiplication between the input layer and *wxh* (because there is no hidden layer value in the previous time step):

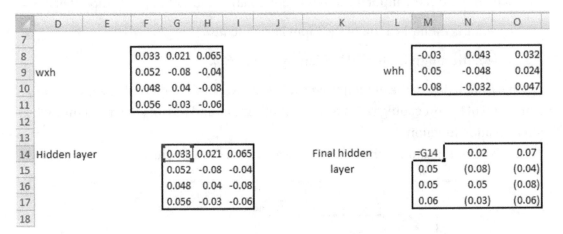

Time Step 2

Starting the second input, the hidden layer consists of the hidden layer component of the current time step and the hidden layer component coming from the previous time step:

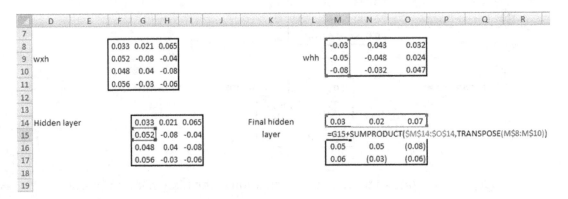

wxh	0.033	0.021	0.065		whh	-0.03	0.043	0.032	
	0.052	-0.08	-0.04			-0.05	-0.048	0.024	
	0.048	0.04	-0.08			-0.08	-0.032	0.047	
	0.056	-0.03	-0.06						
Hidden layer	0.033	0.021	0.065	Final hidden	0.03	0.02	0.07
	0.052	-0.08	-0.04	layer	0.05	=H15+SUMPRODUCT($M$14:$O$14,TRANSPOSE(N$8:N$10))
	0.048	0.04	-0.08		0.05	0.05	(0.08)
	0.056	-0.03	-0.06		0.06	(0.03)	(0.06)

Time Step 3

Similarly, at the third time step, the inputs would be the input at the current time step and the hidden unit values coming from the previous time step. Note that the hidden unit in the previous time step (t-1) is influenced by the hidden values coming from (t-2) also.

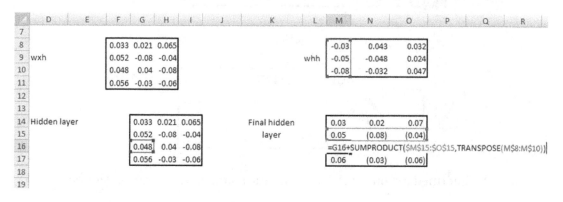

Similarly, the hidden layer values are calculated at the fourth time step.

Now that we have our hidden layer calculated, we pass it through an activation, just as we did it in traditional NN:

Final hidden layer				Tanh activation		
0.03	0.02	0.07		0.03	0.02	0.06
0.05	(0.08)	(0.04)		0.05	-0.08	-0.04
0.05	0.05	(0.08)		0.05	0.05	-0.08
0.06	(0.03)	(0.06)		0.06	-0.03	-0.06

Given that the output from hidden layer activation is 1×3 in size for each input, in order to get an output of 1×4 in size (as the one-hot-encoded version of the expected output "example" is 4 columns in size), the hidden layer *why* should be 3×4 in dimension:

	Q	R	S	T	U
11					
12		Tanh Activation			
13					
14	0.03	0.02	0.06		
15	0.05	-0.08	-0.04		
16	0.05	0.05	-0.08		
17	0.06	-0.03	-0.06		
18					
19		Why			
20					
21	0.058	-0.048	-0.008	0.045	
22	0.007	0.053	-0.092	-0.035	
23	-0.076	-0.072	-0.066	-0.004	
24					
25		Intermediate output			
26	=SUMPRODUCT($Q14:$S14,TRANSPOSE(Q$21:Q$23))				
27	0.00	0.00	0.01	0.01	
28	0.01	0.01	0.00	0.00	
29	0.01	0.00	0.01	0.00	

From the intermediate output, we perform the softmax activation as follows:

	Q	R	S	T
24				
25		Intermediate output		
26	0.00	-0.01	-0.01	0.00
27	0.00	0.00	0.01	0.01
28	0.01	0.01	0.00	0.00
29	0.01	0.00	0.01	0.00
30				
31		Softmax step 1		
32	=EXP(Q26)		0.99	1.00
33	1.00	1.00	1.01	1.01
34	1.01	1.01	1.00	1.00
35	1.01	1.00	1.01	1.00

The second step of softmax would be to normalize each cell value to obtain a probability value:

�F	Q	R	S	T
30				
31		Softmax step 1		
32	1.00	0.99	0.99	1.00
33	1.00	1.00	1.01	1.01
34	1.01	1.01	1.00	1.00
35	1.01	1.00	1.01	1.00
36				
37		Softmax step 2		
38	=Q32/SUM($Q32:$T32)			0.250999
39	0.250214	0.248029	0.25145	0.250307
40	0.251327	0.250374	0.249082	0.249217
41	0.250844	0.248882	0.250469	0.249805

Once the probabilities are obtained, the loss is calculated by taking the cross entropy loss between the prediction and actual output.

Finally, we will be minimizing the loss through the combination of forward and backward propagation epochs in a similar manner as that of NN.

Implementing RNN: SimpleRNN

To see how RNN is implemented in keras, let's go through a simplistic example (only to understand the keras implementation of RNN and then to solidify our understanding by implementing in Excel): classifying two sentences (which have an exhaustive list of three words). Through this toy example, we should be in a better position to understand the outputs quickly (code available as "simpleRNN.ipynb" in github):

```
from keras.preprocessing.text import one_hot
from keras.preprocessing.sequence import pad_sequences
from keras.models import Sequential
from keras.layers import Dense
from keras.layers import Flatten
from keras.layers.recurrent import SimpleRNN
from keras.layers.embeddings import Embedding
from keras.layers import LSTM
import numpy as np
```

Initialize the documents and encode the words corresponding to those documents:

```
# define documents
docs = ['very good',
            'very bad']
# define class labels
labels = [1,0]
from collections import Counter
counts = Counter()
for i,review in enumerate(docs):
    counts.update(review.split())
words = sorted(counts, key=counts.get, reverse=True)
vocab_size=len(words)
word_to_int = {word: i for i, word in enumerate(words, 1)}
encoded_docs = []
for doc in docs:
    encoded_docs.append([word_to_int[word] for word in doc.split()])
```

Pad the documents to a maximum length of two words—this is to maintain consistency so that all the inputs are of the same size:

```
# pad documents to a max length of 2 words
max_length = 2
padded_docs = pad_sequences(encoded_docs, maxlen=max_length, padding='pre')
print(padded_docs)
```

$$[[1\ 3]$$
$$[1\ 2]]$$

Compiling a Model

The input shape to the SimpleRNN function should be of the form (number of time steps, number of features per time step). Also, in general RNN uses tanh as the activation function. The following code specifies the input shape as (2,1) because each input is based on two time steps and each time step has only one column representing it. unroll=True indicates that we are considering previous time steps:

```
# define the model
embed_length=1
```

```
max_length=2
model = Sequential()
model.add(SimpleRNN(1,activation='tanh', return_sequences=False,recurrent_
initializer='Zeros',input_shape=(max_length,embed_length),unroll=True))
model.add(Dense(1, activation='sigmoid'))
# compile the model
model.compile(optimizer='adam', loss='binary_crossentropy', metrics=['acc'])
# summarize the model
print(model.summary())
```

SimpleRNN(1,) indicates that there is a neuron in the hidden layer. return_sequences is false because we are not returning any sequence of outputs, and it is a single output:

Layer (type)	Output Shape	Param #
simple_rnn_16 (SimpleRNN)	(None, 1)	3
dense_12 (Dense)	(None, 1)	2

```
Total params: 5
Trainable params: 5
Non-trainable params: 0
```

Once the model is compiled, let's go ahead and fit the model, as follows:

```
model.fit(padded_docs.reshape(2,2,1),np.array(labels).reshape(max_
length,1),epochs=500)
```

```
Epoch 495/500
2/2 [==============================] - 0s 6ms/step - loss: 0.6112 - acc: 1.0000
Epoch 496/500
2/2 [==============================] - 0s 6ms/step - loss: 0.6108 - acc: 1.0000
Epoch 497/500
2/2 [==============================] - 0s 7ms/step - loss: 0.6104 - acc: 1.0000
Epoch 498/500
2/2 [==============================] - 0s 7ms/step - loss: 0.6100 - acc: 1.0000
Epoch 499/500
2/2 [==============================] - 0s 4ms/step - loss: 0.6097 - acc: 1.0000
Epoch 500/500
2/2 [==============================] - 0s 5ms/step - loss: 0.6093 - acc: 1.0000
```

Note that we have reshaped padded_docs. That's because we need to convert our training dataset into a format as follows while fitting: {data size, number of time steps, features per time step}. Also, labels should be in an array format, since the final dense layer in the compiled model expects an array.

Verifying the Output of RNN

Now that we have fit our toy model, let's verify the Excel calculations we created earlier. Note that we have taken the input to be the raw encodings {1,2,3}—in practice we would never take the raw encodings as they are, but would one-hot-encode or create embeddings for the input. We are taking the raw inputs as they are in this section only to compare the outputs from keras and the hand calculations we are going to do in Excel.

model.layers specifies the layers in the model, and weights gives us an understanding of the layers associated with the model:

```
[<keras.layers.recurrent.SimpleRNN at 0x7fc76516d940>,
 <keras.layers.core.Dense at 0x7fc76516d710>]
```

model.weights gives us an indication of the names associated with the weights in the model:

```
[<tf.Variable 'simple_rnn_2/kernel:0' shape=(1, 1) dtype=float32_ref>,
 <tf.Variable 'simple_rnn_2/recurrent_kernel:0' shape=(1, 1) dtype=float32_ref>,
 <tf.Variable 'simple_rnn_2/bias:0' shape=(1,) dtype=float32_ref>,
 <tf.Variable 'dense_2/kernel:0' shape=(1, 1) dtype=float32_ref>,
 <tf.Variable 'dense_2/bias:0' shape=(1,) dtype=float32_ref>]
```

model.get_weights() gives us the actual values of weights associated with the model:

```
[array([[0.56373304]], dtype=float32),
 array([[-0.50989217]], dtype=float32),
 array([-0.69803804], dtype=float32),
 array([[0.5092683]], dtype=float32),
 array([-0.27220774], dtype=float32)]
```

Note that the weights are ordered—that is, the first weight value corresponds to kernel:0. In other words, it is the same as *wxh*, which is the weight associated with the inputs.

recurrent_kernel:0 is the same as *whh*, which is the weight associated with the connection between the previous hidden layer earlier and the current time step's hidden layer. bias:0 is the bias associated with the inputs. dense_2/kernel:0 is *why*—that is, the weight connecting the hidden layer to the output. dense_2/bias:0 is the bias associated with connection between the hidden layer and the output.

Let's verify the prediction for the input [1,3]:

```
padded_docs[0].reshape(1,2,1)
```

```
array([[[1],
        [3]]], dtype=int32)
```

```
import numpy as np
model.predict(padded_docs[0].reshape(1,2,1))
```

```
array([[0.53199273]], dtype=float32)
```

Given that the prediction is 0.53199 for the inputs [1,3] (in that order), let's verify the same in Excel (available as "simple RNN working verification.xlsx" in github):

Wxh	0.563733
Whh	-0.50989
bx	-0.69804
Why	0.509268
by	-0.27221

The input value at the two time steps are as follows:

		Time step	
		0	1
Wxh	0.563733	Input value	1	3
Whh	-0.50989			
bx	-0.69804			
Why	0.509268			
by	-0.27221			

The matrix multiplication between inputs and weights is calculated as follows:

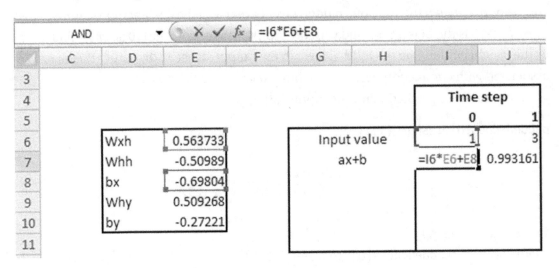

Now that the matrix multiplication is done, we will go ahead and calculate the hidden layer value in time step 0:

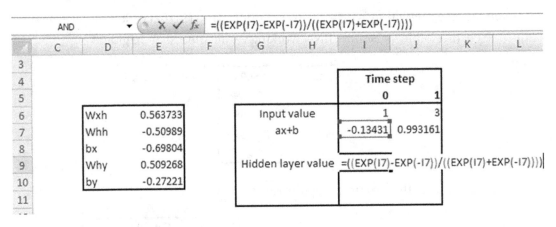

The hidden layer value in time step 1 is going to be the following:

tanh(Hidden layer value in time step 1 × Weight associated with hidden layer to hidden layer connection (whh) + Previous hidden layer value)

Let's calculate the inner part of the tanh function first:

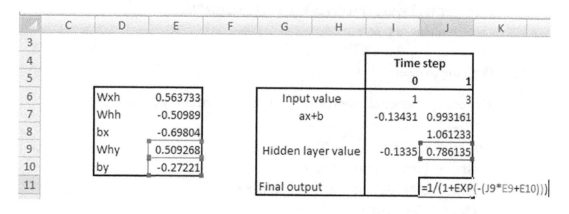

	C	D	E	F	G	H	I	J
3								
4							Time step	
5							0	1
6		Wxh	0.563733		Input value		1	3
7		Whh	-0.50989		ax+b		-0.13431	0.993161
8		bx	-0.69804					=I9*E7+J7
9		Why	0.509268		Hidden layer value		-0.1335	
10		by	-0.27221					
11								

Now we'll calculate the final hidden layer value of time step 1:

	C	D	E	F	G	H	I	J	K	L	M
3											
4							Time step				
5							0	1			
6		Wxh	0.563733		Input value		1	3			
7		Whh	-0.50989		ax+b		-0.13431	0.993161			
8		bx	-0.69804					1.061233			
9		Why	0.509268		Hidden layer value		-0.1335	=(EXP(J8)-EXP(-J8))/(EXP(J8)+EXP(-J8))			
10		by	-0.27221								
11											

Once the final hidden layer value is calculated, it is passed through a sigmoid layer, so the final output is calculated as follows:

	C	D	E	F	G	H	I	J	K
3									
4							Time step		
5							0	1	
6		Wxh	0.563733		Input value		1	3	
7		Whh	-0.50989		ax+b		-0.13431	0.993161	
8		bx	-0.69804					1.061233	
9		Why	0.509268		Hidden layer value		-0.1335	0.786135	
10		by	-0.27221						
11						Final output		=1/(1+EXP(-(J9*E9+E10)))	

The final output that we have from Excel is the same as what we got from keras as output and thus is a verification of the formulas that we looked at earlier:

	C	D	E	F	G	H	I	J
3								
4						Time step		
5							0	1
6		Wxh	0.563733		Input value		1	3
7		Whh	-0.50989		ax+b		-0.13431	0.993161
8		bx	-0.69804					1.061233
9		Why	0.509268		Hidden layer value		-0.1335	0.786135
10		by	-0.27221					
11					Final output			0.531993

Implementing RNN: Text Generation

Now that we've seen how a typical RNN works, let's look into how to generate text using APIs provided by keras for RNN (available as "RNN text generation.ipynb" in github).

For this example, we will be working on the alice dataset (www.gutenberg.org/ebooks/11):

1. Import the packages:

```
from keras.models import Sequential
from keras.layers import Dense,Activation
from keras.layers.recurrent import SimpleRNN
import numpy as np
```

2. Read the dataset:

```
fin=open('/home/akishore/alice.txt',encoding='utf-8-sig')
lines=[]
for line in fin:
  line = line.strip().lower()
  line = line.decode("ascii","ignore")
  if(len(line)==0):
    continue
  lines.append(line)
```

```
fin.close()
text = " ".join(lines)
```

3. Normalize the file to have only small case and remove punctuation, if any:

```
text[:100]
```

u'alice was beginning to get very tired of sitting by her sister on the bank, and of having nothing to'

```
# Remove punctuations in dataset
import re
text = text.lower()
text = re.sub('[^0-9a-zA-Z]+',' ',text)
```

4. One-hot-encode the words:

```
from collections import Counter
counts = Counter()
counts.update(text.split())
words = sorted(counts, key=counts.get, reverse=True)
chars = words
total_chars = len(set(chars))
nb_chars = len(text.split())
char2index = {word: i for i, word in enumerate(chars)}
index2char = {i: word for i, word in enumerate(chars)}
```

5. Create the input and target datasets:

```
SEQLEN = 10
STEP = 1
input_chars = []
label_chars = []
text2=text.split()
for i in range(0,nb_chars-SEQLEN,STEP):
    x=text2[i:(i+SEQLEN)]
    y=text2[i+SEQLEN]
    input_chars.append(x)
```

```
        label_chars.append(y)
    print(input_chars[0])
    print(label_chars[0])
```

```
[u'alice', u'was', u'beginning', u'to', u'get', u'very', u'tired', u'of', u'sitting', u'by']
her
```

6. Encode the input and output datasets:

```
X = np.zeros((len(input_chars), SEQLEN, total_chars), dtype=np.bool)
y = np.zeros((len(input_chars), total_chars), dtype=np.bool)
# Create encoded vectors for the input and output values
for i, input_char in enumerate(input_chars):
    for j, ch in enumerate(input_char):
        X[i, j, char2index[ch]] = 1
    y[i,char2index[label_chars[i]]]=1
print(X.shape)
print(y.shape)
```

```
        (30407, 10, 3028)
        (30407, 3028)
```

Note that, the shape of X indicates that we have a total 30,407 rows
that have 10 words each, where each of the 10 words is expressed
in a 3,028-dimensional space (since there are a total of 3,028
unique words).

7. Build the model:

```
HIDDEN_SIZE = 128
BATCH_SIZE = 128
NUM_ITERATIONS = 100
NUM_EPOCHS_PER_ITERATION = 1
NUM_PREDS_PER_EPOCH = 100
model = Sequential()
model.add(SimpleRNN(HIDDEN_SIZE,return_sequences=False,input_
shape=(SEQLEN,total_chars),unroll=True))
model.add(Dense(nb_chars, activation='sigmoid'))
```

```
model.compile(optimizer='rmsprop', loss='categorical_crossentropy')
model.summary()
```

```
Layer (type)                  Output Shape            Param #
=================================================================
simple_rnn_1 (SimpleRNN)      (None, 128)             404096
_____
dense_1 (Dense)               (None, 3028)            390612
=================================================================
Total params: 794,708
Trainable params: 794,708
Non-trainable params: 0
_____
```

8. Run the model, where we randomly generate a seed text and try to predict the next word given the set of seed words:

```
for iteration in range(150):
    print("=" * 50)
    print("Iteration #: %d" % (iteration))
    # Fitting the values
    model.fit(X, y, batch_size=BATCH_SIZE, epochs=NUM_EPOCHS_PER_
    ITERATION)

    # Time to see how our predictions fare
    # We are creating a test set from a random location in our dataset
    # In the code below, we are selecting a random input as our
      seed value of words
    test_idx = np.random.randint(len(input_chars))
    test_chars = input_chars[test_idx]
    print("Generating from seed: %s" % (test_chars))
    print(test_chars)
    # From the seed words, we are tasked to predict the next words
    # In the code below, we are predicting the next 100 words
      (NUM_PREDS_PER_EPOCH) after the seed words
    for i in range(NUM_PREDS_PER_EPOCH):
        # Pre processing the input data, just like the way we did
          before training the model
        Xtest = np.zeros((1, SEQLEN, total_chars))
```

237

```
for i, ch in enumerate(test_chars):
    Xtest[0, i, char2index[ch]] = 1
# Predict the next word
pred = model.predict(Xtest, verbose=0)[0]
# Given that, the predictions are probability values, we take
    the argmax to fetch the location of highest probability
# Extract the word belonging to argmax
ypred = index2char[np.argmax(pred)]
print(ypred,end=' ')
# move forward with test_chars + ypred so that we use the
    original 9 words + prediction for the next prediction
test_chars = test_chars[1:] + [ypred]
```

The output in the initial iterations is just the single word *the*—always!

The output at the end of 150 iterations is as follows (note that the below is only a partial output):

```
Epoch 1/1
30407/30407 [==============================] - 3s 97us/step - loss: 0.9459
Generating from seed: ['and', 'i', 'm', 'i', 'and', 'oh', 'dear', 'how', 'puzzli
['and', 'i', 'm', 'i', 'and', 'oh', 'dear', 'how', 'puzzling', 'it']
again said the mock turtle said the duchess looked nearly on a lobsters of verse
```

The preceding output has very little loss. And if you look at the output carefully after you execute code, after some iterations it is reproducing the exact text that is present in the dataset—a potential overfitting issue. Also, notice the shape of our input: ~30K inputs, where there are 3,028 columns. Given the low ratio of rows to columns, there is a chance of overfitting. This is likely to work better as the number of input samples increases a lot more.

The issue of having a high number of columns can be overcome by using embedding, which is very similar to the way in which we calculated word vectors. Essentially, *embeddings* represent a word in a much lower dimensional space.

Embedding Layer in RNN

To see how embedding works, let's look at a dataset that tries to predict customer sentiment of an airline based on customer tweets (code available as "RNNsentiment. ipynb" in github):

1. As always, import the relevant packages:

```
#import relevant packages
from keras.layers import Dense, Activation
from keras.layers.recurrent import SimpleRNN
from keras.models import Sequential
from keras.utils import to_categorical
from keras.layers.embeddings import Embedding
from sklearn.cross_validation import train_test_split
import numpy as np
import nltk
from nltk.corpus import stopwords
import re
import pandas as pd
#Let us go ahead and read the dataset:
t=pd.read_csv('/home/akishore/airline_sentiment.csv')
t.head()
```

	airline_sentiment	text
0	positive	@VirginAmerica plus you've added commercials t...
1	negative	@VirginAmerica it's really aggressive to blast...
2	negative	@VirginAmerica and it's a really big bad thing...
3	negative	@VirginAmerica seriously would pay $30 a fligh...
4	positive	@VirginAmerica yes, nearly every time I fly VX...

```
import numpy as np
t['sentiment']=np.where(t['airline_sentiment']=="positive",1,0)
```

2. Given that the text is noisy, we will pre-process it by removing punctuation and also converting all words into lowercase:

```
def preprocess(text):
    text=text.lower()
    text=re.sub('[^0-9a-zA-Z]+',' ',text)
    words = text.split()
    #words2=[w for w in words if (w not in stop)]
    #words3=[ps.stem(w) for w in words]
    words4=' '.join(words)
    return(words4)
t['text'] = t['text'].apply(preprocess)
```

3. Similar to how we developed in the previous section, we convert each word into an index value as follows:

```
from collections import Counter
counts = Counter()
for i,review in enumerate(t['text']):
    counts.update(review.split())
words = sorted(counts, key=counts.get, reverse=True)
words[:10]
```

```
['to', 'i', 'the', 'a', 'you', 'united', 'for', 'flight', 'and', 'on']
```

```
chars = words
nb_chars = len(words)
word_to_int = {word: i for i, word in enumerate(words, 1)}
int_to_word = {i: word for i, word in enumerate(words, 1)}
word_to_int['the']
#3
int_to_word[3]
#the
```

4. Map each word in a review to its corresponding index:

```
mapped_reviews = []
for review in t['text']:
    mapped_reviews.append([word_to_int[word] for word in review.
split()])
t.loc[0:1]['text']
```

```
    0     virginamerica plus you ve added commercials to...
    1     virginamerica it s really aggressive to blast ...
Name: text, dtype: object
```

```
mapped_reviews[0:2]
```

```
[[104, 575, 5, 84, 1320, 2497, 1, 3, 179, 7250],
 [104,
  17,
  32,
  124,
  3331,
  1,
  4219,
  5487,
  959,
  16,
  22,
  3296,
  5273,
  62,
  52,
  27,
  479,
  2521]]
```

Note that, the index of virginamerica is the same in both reviews (104).

5. Initialize a sequence of zeroes of length 200. Note that we have chosen 200 as the sequence length because no review has more than 200 words in it. Moreover, the second part of the following code makes sure that for all reviews that are less than 200 words in size, all the starting indices are zero padded and only the last indices are filled with index corresponding to the words present in the review:

```
sequence_length = 200
sequences = np.zeros((len(mapped_reviews), sequence_
length),dtype=int)
for i, row in enumerate(mapped_reviews):
    review_arr = np.array(row)
    sequences[i, -len(row):] = review_arr[-sequence_length:]
```

6. We further split the dataset into train and test datasets, as follows:

```
y=t['sentiment'].values
X_train, X_test, y_train, y_test = train_test_split(sequences, y,
test_size=0.30,random_state=10)
y_train2 = to_categorical(y_train)
y_test2 = to_categorical(y_test)
```

7. Once the datasets are in place, we go ahead and create our model, as follows. Note that embedding as a function takes in as input the total number of unique words, the reduced dimension in which we express a given word, and the number of words in an input:

```
top_words=12679
embedding_vecor_length=32
max_review_length=200
model = Sequential()
model.add(Embedding(top_words, embedding_vecor_length,
input_length=max_review_length))
model.add(SimpleRNN(1, return_sequences=False,unroll=True))
model.add(Dense(2, activation='softmax'))
```

```
model.compile(loss='categorical_crossentropy', optimizer='adam',
metrics=['accuracy'])
print(model.summary())
model.fit(X_train, y_train2, validation_data=(X_test, y_test2),
epochs=50, batch_size=1024)
```

Layer (type)	Output Shape	Param #
embedding_14 (Embedding)	(None, 200, 32)	405728
simple_rnn_14 (SimpleRNN)	(None, 1)	34
dense_14 (Dense)	(None, 2)	4

```
Total params: 405,766
Trainable params: 405,766
Non-trainable params: 0
```

Now let's look at the summary output of the preceding model. There are a total of 12,679 unique words in the dataset. The embedding layer ensures that we represent each of the words in a 32-dimensional space, hence the 405,728 parameters in the embedding layer.

Now that we have 32 *embedded dimensional inputs*, each input is now connected to one hidden layer unit—thus 32 weights. Along, with the 32 weights, we would have a bias. The final weight corresponding to this layer would be the weight that connects the previous hidden to unit value to the current hidden unit. Thus a total of 34 weights.

Note that, given that there is an output coming from the embedding layer, we don't need to specify the input shape in the SimpleRNN layer. Once the model has run, the output classification accuracy turns out to be close to 87%.

Issues with Traditional RNN

A traditional RNN that takes multiple time steps into account for giving a prediction is shown in Figure 10-5.

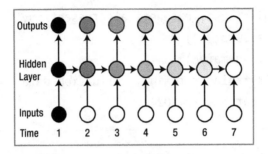

Figure 10-5. *An RNN with multiple time steps*

Note that as time step increases, the impact of input from a much earlier layer on output of later layers is much less. That can be seen in the following (for now, we'll ignore the bias terms):

$$h_1 = Wx_1$$

$$h_2 = Wx_2 + Uh_1 = Wx_2 + UWx_1$$

$$h_3 = Wx_3 + Uh_2 = Wx_3 + UWx_2 + U^2Wx_1$$

$$h_4 = Wx_4 + Uh_3 = Wx_4 + UWX_3 + U^2WX_2 + U^3WX_1$$

$$h_5 = Wx_5 + Uh_4 = Wx_5 + UWX_4 + U^2WX_3 + U^3WX_2 + U^4WX_1$$

Note that as the time stamp increases, the value of the hidden layer is highly dependent on X_1 if $U > 1$, and a little dependent on X_1 if $U < 1$.

The Problem of Vanishing Gradient

The gradient of U^4 with respect to U is $4 \times U^3$. In such a case, note that if $U < 1$, the gradient is very small, so arriving at the ideal weights takes a very long time if the output at a much a later time step depends on the input at a given time step. This results in an issue when there is a dependency on a word that occurred much earlier in the time steps in some sentences. For example, "I am from India. I speak fluent ____." In this case, if we did not take the first sentence into account, the output of the second sentence, "I speak fluent ____" could be the name of any language. Because we mentioned the country in the first sentence, we should be able to narrow things down to languages specific to India.

The Problem of Exploding Gradients

In the preceding scenario, if $U > 1$, then gradients increase by a much larger amount. This would result in having a very high weightage for inputs that occurred much earlier in the time steps and low weightage for inputs that occurred near the word that we are trying to predict.

Hence, depending on the value of U (weights of the hidden layer), the weights either get updated very quickly or take a very long time.

Given that vanishing/exploding gradient is an issue, we should deal with RNNs in a slightly different way.

LSTM

Long short-term memory (LSTM) is an architecture that helps overcome the vanishing or exploding gradient problem we saw earlier. In this section, we will look at the architecture of LSTM and see how it helps in overcoming the issue with traditional RNN.

LSTM is shown in Figure 10-6.

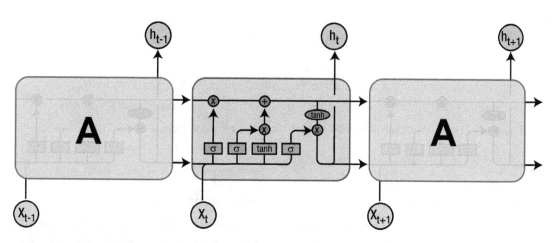

Figure 10-6. *LSTM*

Note that although the input X and the output of the hidden layer (h) remain the same, the activations that happen within the hidden layer are different. Unlike the traditional RNN, which has tanh activation, there are different activations that happen within LSTM. We'll go through each of them.

In Figure 10-7, X and h represent the input and hidden layer, as we saw earlier.

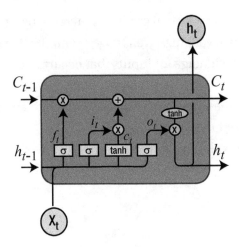

Figure 10-7. *Various components of LSTM*

C represents the cell state. You can think of *cell state* as a way in which long-term dependencies are captured.

f represents the forget gate:

$$f_t = \sigma\left(W_{xf}x^{(t)} + W_{hf}h^{(t-1)} + b_f\right)$$

Note that the sigmoid gives us a mechanism to specify what needs to be forgotten. This way, some historical words that are captured in $h^{(t-1)}$ are selectively forgotten.

Once we figure what needs to be forgotten, the cell state gets updated as follows:

$$c_t = \left(c_{t-1} \otimes f\right)$$

Note that $\otimes$ represents element-to-element multiplication.

Consider that once we fill in the blank in "I live in India. I speak ____" with the name of an Indian language, we don't need the context of "I live in India" anymore. This is where the forget gate helps in selectively forgetting the information that is not needed anymore.

Once we figure out what needs to be forgotten in the cell state, we can go ahead and update the cell state based on the current input.

In the next step, the input that needs to update the cell state is achieved through the sigmoid application on top of input, and the magnitude of update (either positive or negative) is obtained through the tanh activation.

The input can be specified as follows:

$$i_t = \sigma\left(W_{xi}x^{(t)} + W_{hi}h^{(t-1)} + b_i\right)$$

The modulation can be specified like this:

$$g_t = \tanh\left(W_{xg}x^{(t)} + W_{hg}h^{(t-1)} + b_g\right)$$

The cell state thus finally gets updated as the following:

$$C^{(t)} = \left(C^{(t-1)} \odot f_t\right) \oplus \left(i_t \odot g_t\right)$$

In the final gate, we need to specify what part of the combination of input and cell state needs to be outputted to the next hidden layer:

$$o_t = \sigma\left(W_{xo}x^{(t)} + W_{ho}h^{(t-1)} + b_o\right)$$

The final hidden layer is represented like this:

$$h^{(t)} = o_t \odot \tanh\left(C^{(t)}\right)$$

Given that the cell state can memorize the values that are needed at a later point in time, LSTM provides better results than traditional RNN in predicting the next word, typically in sentiment classification. This is especially useful in a scenario where there is a long-term dependency that needs to be taken care of.

Implementing Basic LSTM in keras

To see how the theory presented so far translates into action, let's relook at the toy example we saw earlier (code available as "LSTM toy example.ipynb" in github):

1. Import the relevant packages:

```
from keras.preprocessing.text import one_hot
from keras.preprocessing.sequence import pad_sequences
from keras.models import Sequential
from keras.layers import Dense
```

```
from keras.layers import Flatten
from keras.layers.recurrent import SimpleRNN
from keras.layers.embeddings import Embedding
from keras.layers import LSTM
import numpy as np
```

2. Define documents and labels:

```
# define documents
docs = ['very good',
            'very bad']
# define class labels
labels = [1,0]
```

3. One-hot-encode the documents:

```
from collections import Counter
counts = Counter()
for i,review in enumerate(docs):
    counts.update(review.split())
words = sorted(counts, key=counts.get, reverse=True)
vocab_size=len(words)
word_to_int = {word: i for i, word in enumerate(words, 1)}
encoded_docs = []
for doc in docs:
    encoded_docs.append([word_to_int[word] for word in doc.split()])
encoded_docs
```

```
[[1, 3], [1, 2]]
```

4. Pad documents to a maximum length of two words:

```
max_length = 2
padded_docs = pad_sequences(encoded_docs, maxlen=max_length,
padding='pre')
print(padded_docs)
```

```
[[1 3]
 [1 2]]
```

5. Build the model:

```
model = Sequential()
model.add(LSTM(1,activation='tanh', return_
sequences=False,recurrent_initializer='Zeros',
recurrent_activation='sigmoid',
                input_shape=(2,1),unroll=True))
model.add(Dense(1, activation='sigmoid'))
model.compile(optimizer='adam', loss='binary_crossentropy',
metrics=['acc'])
print(model.summary())
```

Layer (type)	Output Shape	Param #
lstm_10 (LSTM)	(None, 1)	12
dense_10 (Dense)	(None, 1)	2

```
Total params: 14
Trainable params: 14
Non-trainable params: 0
```

Note that in the preceding code, we have initialized the recurrent initializer and
recurrent activation to certain values only to make this toy example easier to understand
when implemented in Excel. The purpose is to help you understand what is happening
in the back end only.

Once the model is initialized as discussed, let's go ahead and fit the model:

```
model.fit(padded_docs.reshape(2,2,1),np.array(labels).reshape(max_
length,1),epochs=500)
```

```
Epoch 496/500
2/2 [==============================] - 0s 3ms/step - loss: 0.6616 - acc: 1.0000
Epoch 497/500
2/2 [==============================] - 0s 3ms/step - loss: 0.6615 - acc: 1.0000
Epoch 498/500
2/2 [==============================] - 0s 3ms/step - loss: 0.6614 - acc: 1.0000
Epoch 499/500
2/2 [==============================] - 0s 3ms/step - loss: 0.6612 - acc: 1.0000
Epoch 500/500
2/2 [==============================] - 0s 3ms/step - loss: 0.6611 - acc: 1.0000
```

The layers of this model are as follows. Here is model.layers:

```
[<keras.layers.recurrent.LSTM at 0x7f56a17930b8>,
 <keras.layers.core.Dense at 0x7f56a1793128>]
```

The weights and the order of weights can be obtained as follows:

model.layers[0].get_weights()

```
[array([[0.47091758, 0.05323942, 0.27755383, 0.9501942 ]], dtype=float32),
 array([[ 0.42902616,  0.15900111, -0.2862077 , -0.6242878 ]],
       dtype=float32),
 array([ 0.19718488,  1.0436496 , -0.21873292, -0.71010154], dtype=float32)]
```

model.layers[0].trainable_weights

```
[<tf.Variable 'lstm_1/kernel:0' shape=(1, 4) dtype=float32_ref>,
 <tf.Variable 'lstm_1/recurrent_kernel:0' shape=(1, 4) dtype=float32_ref>,
 <tf.Variable 'lstm_1/bias:0' shape=(4,) dtype=float32_ref>]
```

model.layers[1].get_weights()

```
[array([[1.4696432]], dtype=float32), array([-0.38569826], dtype=float32)]
```

model.layers[1].trainable_weights

```
[<tf.Variable 'dense_1/kernel:0' shape=(1, 1) dtype=float32_ref>,
 <tf.Variable 'dense_1/bias:0' shape=(1,) dtype=float32_ref>]
```

From the preceding code, we can see that weights of input (kernel) are obtained first, followed by weights corresponding to the hidden layer (recurrent_kernel) and finally the bias in the LSTM layer.

Similarly, in the dense layer (the layer connecting the hidden layer to output), the weight to be multiplied with the hidden layer comes first, followed by the bias.

Also note that the order in which weights and bias appear in the LSTM layer is as follows:

1. Input gate

2. Forget gate

3. Modulation gate (cell gate)

4. Output gate

Now that we have our outputs, let's go ahead and calculate the predictions for input. Note that just like in the previous section, we are using raw encoded inputs (1,2,3) without further processing them—only to see how the calculation works.

In practice, we would be further processing the inputs, potentially encoding them into vectors to obtain the predictions, but in this example we are interested in solidifying our knowledge of how LSTM works by replicating the predictions from LSTM in Excel:

```
padded_docs[1].reshape(1,2,1)
```

```
array([[[1],
        [2]]], dtype=int32)
```

```
model.predict(padded_docs[1].reshape(1,2,1))
```

```
array([[0.44851267]], dtype=float32)
```

Now that we have a predicted probability of 0.4485 from the model, let's hand-calculate the values in Excel (available in github as "LSTM working details.xlsx"):

Gate	weight	recurrent	bias
input	1.1860023	0.06184402	−0.3049411
forget	−0.35046005	0.20871657	0.99633026
cell/ modulation	−0.40737563	0.12195425	0.61977786
output	0.51597786	0.09282562	0.609171

Note that the values here are taken from keras's model.layers[0].get_weights() output.

Before proceeding with the calculation of the values at various gates, note that we have initialized the value of recurrent layer (h_{t-1}) to 0. In the first time step, the input is a value of 1. Let's calculate the value at various gates:

Input	1
cell state0	0
forget1	0.64587021
forget2	0.656079221
cell state1	0
input1	0.88106124
input2	0.707042088
cell1	0.21240223
cell2	0.209264683
cell state2	0.147958938
cell state3	0.147958938
output1	1.12514886
output2	0.754942528

The calculations to obtain the preceding output are as follows:

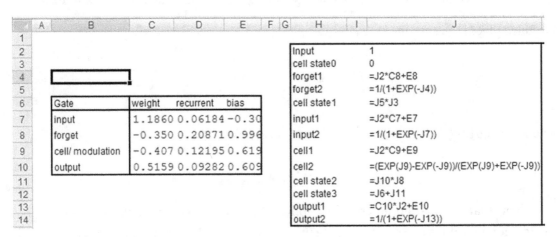

Gate	weight	recurrent	bias
input	1.1860	0.06184	-0.30
forget	-0.350	0.20871	0.996
cell/ modulation	-0.407	0.12195	0.619
output	0.5159	0.09282	0.609

Input	1
cell state0	0
forget1	=J2*C8+E8
forget2	=1/(1+EXP(-J4))
cell state1	=J5*J3
input1	=J2*C7+E7
input2	=1/(1+EXP(-J7))
cell1	=J2*C9+E9
cell2	=(EXP(J9)-EXP(-J9))/(EXP(J9)+EXP(-J9))
cell state2	=J10*J8
cell state3	=J6+J11
output1	=C10*J2+E10
output2	=1/(1+EXP(-J13))

Now that all the values at various gates are calculated, we'll calculate the output (the hidden layer):

	F G	H	I	J
1				
2		Input	1	
3		cell state0	0	
4		forget1	=J2*C8+E8	
5		forget2	=1/(1+EXP(-J4))	
6		cell state1	=J5*J3	
7		input1	=J2*C7+E7	
8		input2	=1/(1+EXP(-J7))	
9		cell1	=J2*C9+E9	
10		cell2	=(EXP(J9)-EXP(-J9))/(EXP(J9)+EXP(-J9))	
11		cell state2	=J10*J8	
12		cell state3	=J6+J11	
13		output1	=C10*J2+E10	
14		output2	=1/(1+EXP(-J13))	
15				
16		hidden layer	=(EXP(J12)-EXP(-J12))/(EXP(J12)+EXP(-J12))*J14	

The hidden layer value just shown is the hidden layer output at the time step where the input is 1.

Now, we'll go ahead and calculate the hidden layer value when the input is 2 (which is the input at the second time step of our data point that we were predicting in the code earlier):

Input	1	2
cell state0	0	0.147958938
forget1	0.64587021	0.318555254
forget2	0.656079221	0.578972116
cell state1	0	0.0856641
input1	0.88106124	2.073921576
input2	0.707042088	0.888342536
cell1	0.21240223	-0.181449593
cell2	0.209264683	-0.179484127
cell state2	0.147958938	-0.159443385
cell state3	0.147958938	-0.073779285
output1	1.12514886	1.651420381
output2	0.754942528	0.839082927
hidden layer	0.11089246	-0.061794855

Let's see how the values are obtained for the various gates and the hidden layer for the second input. The key to note here is that the hidden layer of the first time step output is an input to the calculation of all gates in the second input:

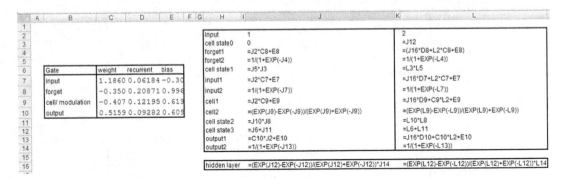

Finally, given that we have calculated the hidden layer output of the second time step, we calculate the output, as follows:

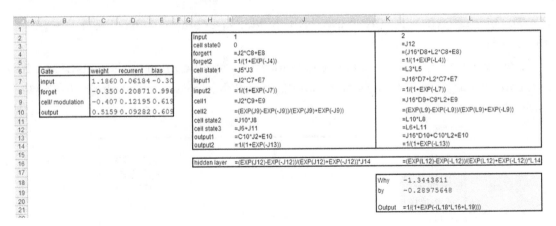

The final output of the preceding calculations is shown here:

Input	1	2
cell state0	0	0.147958938
forget1	0.64587021	0.318555254
forget2	0.656079221	0.578972116
cell state1	0	0.0856641
input1	0.88106124	2.073921576
input2	0.707042088	0.888342536
cell1	0.21240223	-0.181449593
cell2	0.209264683	-0.179484127
cell state2	0.147958938	-0.159443385
cell state3	0.147958938	-0.073779285
output1	1.12514886	1.651420381
output2	0.754942528	0.839082927

hidden layer	0.11089246	-0.061794855

Why	-1.3443611
by	-0.28975648
Output	0.448512683

Note that the output that we've derived is the same as what we see in the keras output.

Implementing LSTM for Sentiment Classification

In the last section, we implemented sentiment classification using RNN in keras. In this section, we will look at implementing the same using LSTM. The only change in the code we saw above will be the model compiling part, where we will be using LSTM in place of SimpleRNN—everything else will remain the same (code is available in "RNN sentiment. ipynb" file in github):

```
top_words=nb_chars
embedding_vecor_length=32
max_review_length=200
model = Sequential()
model.add(Embedding(top_words, embedding_vecor_length, input_length=max_
review_length))
model.add(LSTM(10))
model.add(Dense(2, activation='softmax'))
```

```
model.compile(loss='categorical_crossentropy', optimizer='adam',
metrics=['accuracy'])
print(model.summary())
model.fit(X_train, y_train2, validation_data=(X_test, y_test2), epochs=50,
batch_size=1024)
```

Once you implement the model, you should see that the prediction accuracy of LSTM is slightly better than that of RNN. In practice, for the dataset we looked earlier, LSTM gives an accuracy of 91%, whereas RNN gave an accuracy of 87%. This can be further fine-tuned by adjusting various hyper-parameters that are provided by the functions.

Implementing RNN in R

To look at how to implement RNN/LSTM in R, we will use the IMDB sentiment classification dataset that comes pre-built along with the kerasR package (code available as "kerasR_code_RNN.r" in github):

```
# Load the dataset
library(kerasR)
imdb <- load_imdb(num_words = 500, maxlen = 100)
```

Note that we are fetching only the top 500 words by specifying num_words as a parameter. We are also fetching only those IMDB reviews that have a length of at most 100 words.

Let's explore the structure of the dataset:

```
str(imdb)
```

We should notice that in the pre-built IMDB dataset that came along with the kerasR package, each word is replaced by the index it represents by default. So we do not have to perform the step of word-to-index mapping:

```
# Build the model with an LSTM
model <- Sequential()
```

```
model$add(Embedding(500, 32, input_length = 100, input_shape = c(100)))
```

```
model$add(LSTM(32)) # Use SimpleRNN, if we were to perform a RNN function
model$add(Dense(256))
```

```
model$add(Activation('relu'))
model$add(Dense(1))
model$add(Activation('sigmoid'))
# Compile and fit the model
keras_compile(model, loss = 'binary_crossentropy', optimizer =
Adam(),metrics='binary_accuracy')
keras_fit(model, X_train, Y_train, batch_size = 1024, epochs = 50,
verbose = 1,validation_data = list(X_test,Y_test))
```

The preceding results in close to 79% accuracy on the test dataset prediction.

Summary

In this chapter, you learned the following:

- RNNs are extremely helpful in dealing with data that has time dependency.

- RNNs face issues with vanishing or exploding gradient when dealing with long-term dependency in data.

- LSTM and other recent architectures come in handy in such a scenario.

- LSTM works by storing the information in cell state, forgetting the information that does not help anymore, selecting the information as well as the amount of information that need to be added to cell state based on current input, and finally, the information that needs to be outputted to the next state.

CHAPTER 11

Clustering

The dictionary meaning of clustering is *grouping*. In data science, too, clustering is an unsupervised learning technique that helps in grouping our data points.

The benefits of grouping data points (rows) include the following:

- For a business user to understand the various types of users among customers

- To make business decisions at a cluster (group) level rather than at an overall level

- To help improve the accuracy of predictions, since different groups exhibit different behavior, and hence a separate model can be made for each group

In this chapter, you will learn the following:

- Different types of clustering

- How different types of clustering work

- Use cases of clustering

Intuition of clustering

Let's consider an example of a retail store with 4,000 outlets. The central planning team has to perform year-end evaluations of the store managers of all the outlets. The major metric on which a store manager is evaluated is the total sales made by the store over the year.

- *Scenario 1*: A store manager is evaluated solely on the sales made. We'll rank order all the store managers by the sales their stores have made, and the top-selling store managers receive the highest reward.

© V Kishore Ayyadevara 2018

V. K. Ayyadevara, *Pro Machine Learning Algorithms*, https://doi.org/10.1007/978-1-4842-3564-5_11

Drawback: The major drawback of this approach is that we are not taking into account that some stores are in cities, which typically have very high sales when compared to rural stores. The biggest reason for high sales in city stores could be that the population is high and/or the spending power of city-based customers is higher.

- *Scenario 2*: If we could divide the stores into city and rural, or stores that are frequented by customers with high purchasing power, or stores that are frequented by populations of a certain demographic (say, young families) and call each of them a *cluster*, then only those store managers who belong to the same cluster can be compared.

For example, if we divide all the stores into city stores and rural stores, then all the store managers of city stores can be compared with each other, and the same would be the case with rural store managers.

Drawback: Although we are in a better place to compare the performance of different store managers, the store managers can still be compared unfairly. For example, two city stores could be different—one store is in the central business district frequented by office workers, and another is in a residential area of a city. It is quite likely that central business district stores inherently have higher sales among the city stores.

Although scenario 2 still has a drawback, it is not as bad a problem as in scenario 1. Thus clustering into two kinds of stores helps a little as a more accurate way of measuring store manager performance.

Building Store Clusters for Performance Comparison

In scenario 2, we saw that the store managers can still question the process of comparison as unfair—because stores can still differ in one or more parameters.

The store managers might cite multiple reasons why theirs differs from other stores:

- Differences in products that are sold in different stores

- Differences in the age group of customers who visit the store

- Differences in lifestyles of the customers who visit the store

For now, for the sake of simplicity, let's define all our groups:

- City store versus rural store

- High-products store versus low-products store

- High-age-group-customers store versus low-age-group-customers store

- Premium-shoppers store versus budget-shoppers store

We can create a total of 16 different clusters (groups) based on just the factors listed (the exhaustive list of all combinations results in 16 groups).

We talked about the four important factors that differentiate the stores, but there could still be many other factors that differentiate among stores. For example, a store that is located in a rainier part of the state versus a store that is located in a sunnier part.

In essence, stores differ from each other in multiple dimensions/factors. But results of store managers might differ significantly because of certain factors, while the impact of other factors could be minimal.

Ideal Clustering

So far, we see that each store is unique, and there could be a scenario where a store manager can always cite one reason or another why they cannot be compared with other stores that fall into the same cluster. For example, a store manager can say that although all the stores that belong to this group are city stores with majority of premium customers in the middle-aged category, *their* store did not perform as well as other stores because it is located close to a competitor store that has been promoting heavily—hence, the store sales were not that high when compared to other stores in the same group.

Thus, if we take all the reasons into consideration, we might end up with a granular segmentation to the extent that there is only store in each cluster. That would result in a great clustering output, where each store is different, but the outcome would be useless because now we cannot compare across stores.

Striking a Balance Between No Clustering and Too Much Clustering: K-means Clustering

We've seen that having as many clusters as there are stores is a great cluster that differentiates among each store, but it's a useless cluster since we cannot draw any meaningful results from it. At the same time, having no cluster is also bad because store managers might be compared inaccurately with store mangers of completely different stores.

Thus, using the clustering process, we strive towards striking a balance. We want to identify the few factors that differentiate the stores as much as possible and consider only those for evaluation while leaving out the rest. The means of identifying the few factors that differentiate among the stores as much as possible is achieved using *k-means clustering*.

To see how k-means clustering works, let's consider a scenario: you are the owner of a pizza chain. You have a budget to open three new outlets in a neighborhood. How do you come up with the optimal places to open the three outlets?

For now, we'll assume that the traffic in the neighborhood is uniform across all the roads. Let's say our neighborhood looks like Figure 11-1.

Figure 11-1. *Each marker represents a household*

Ideally, we would have come up with three outlets that are far away from each other but that collectively are nearest to most of the neighborhood. For example, something like Figure 11-2.

Figure 11-2. *The circles represent potential outlet locations*

That seems okay, but can we come up with more optimal outlet locations? Let's try an exercise. We will attempt to do the following:

1. Minimize the distance of each household to the nearest pizza outlet

2. Maximize the distance between each pizza outlet

Assume for a second that a pizza outlet can deliver to two houses only. If the demand from both houses is uniform, would it be better to locate the pizza outlet exactly between the houses or closer to one or the other house? If it takes 15 minutes to deliver to house A and 45 minutes to deliver to house B, intuitively we would seem to be better off locating the outlet where the time to deliver to either house is 30 minutes—that is, at a point exactly between the two houses. If that were not the case, the outlet might often fail on its promise to deliver within 45 minutes for house B, while never failing to keep its promises to household A.

The Process of Clustering

In the scenario underway, how do we come up with a more scientific way of identifying the right pizza delivery outlets? The process, or algorithm, is as follows:

1. Randomly come up with three places where we can start outlets, as shown in Figure 11-3.

Figure 11-3. *Random locations*

2. Measure the distance of each house to the three outlet locations. The outlet that is closest to the household delivers to the household. The scenario would look like Figure 11-4.

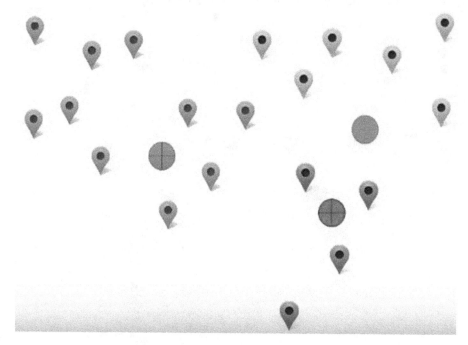

Figure 11-4. *Better informed locations*

3. As we saw earlier, we are better off if the delivery outlet is in the middle of households than far away from the majority of the households. Thus, let's change our earlier planned outlet location to be in the middle of the households (Figure 11-5).

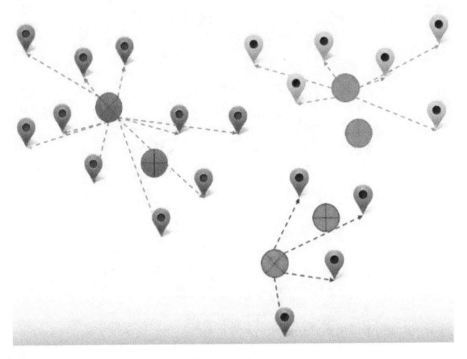

Figure 11-5. *Locations in the middle*

4. We see that the delivery outlet locations have changed to be
    more in the middle of each group. But because of the change
    in location, there might be some households that are now closer
    to a different outlet. Let's reassign households to stores based on
    their distance to different stores (Figure 11-6).

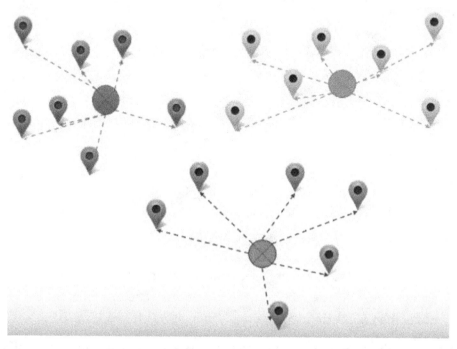

Figure 11-6. *Reassigning households*

5. Now that some households (comparing Figures 11-5
    and 11-6) have a different outlet that serves them, let's recompute
    the middle point of that group of households (Figure 11-7).

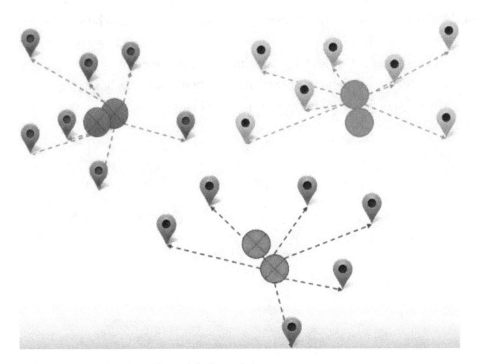

Figure 11-7. *Recomputing the middles*

6. Now that the cluster centers have changed, there is now another
    a chance that the households need to be reassigned to a different
    outlet than the current one.

The steps continue until there is no further reassignment of households to different
clusters, or a maximum of a certain number of iterations is reached.

As you can see, we can in fact come up a more scientific/analytical way of finding
optimal locations where the outlets can be opened.

Working Details of K-means Clustering Algorithm

We will be opening three outlets, so we came up with three groups of households, where
each group is served by a different outlet.

The *k* in *k-means clustering* stands for the number of groups we will be creating in the
dataset we have. In some of the steps in the algorithm we went through in the preceding
section, we kept updating the centers once some households changed their group from
one to another. The way we were updating centers is by taking the *mean* of all the data
points, hence *k-means*.

Finally, after going through the steps as described, we ended up with three groups of three datapoints or three clusters from the original dataset.

Applying the K-means Algorithm on a Dataset

Let's look at how k-means clustering can be implemented on a dataset (available as "clustering process.xlsx" in github), as follows:

X	Y	Cluster
5	0	1
5	2	2
3	1	1
0	4	2
2	1	1
4	2	2
2	2	1
2	3	2
1	3	1
5	4	2

Let X and Y be the independent variables that we want our clusters to be based on. Let's say we want to divide this dataset into two clusters.

In the first step, we initialize the clusters randomly. So, the cluster column in the preceding table is randomly initialized.

Let's calculate the *centroid* of each of the clusters:

Centroid	1	2
X	2.6	3.2
Y	1.4	3

Note that the value 2.6 is the average of all the values of X that belong to cluster 1. Similarly, the other centers are calculated.

Now let's calculate the distance of each point to the two cluster centers. The cluster center that is closest to the data point is the cluster the data point should belong to:

K-means Algorithm

X	Y	Cluster	Centroid	1	2	Dist-sq	1	2	Cluster
5	0	1	X	2.6	3.2	7.72	7.72	12.24	1
5	2	2	Y	1.4	3	4.24	6.12	4.24	2
3	1	1				0.32	0.32	4.04	1
0	4	2				11.24	13.52	11.24	2
2	1	1				0.52	0.52	5.44	1
4	2	2				1.64	2.32	1.64	2
2	2	1				0.72	0.72	2.44	1
2	3	2				1.44	2.92	1.44	2
1	3	1				4.84	5.12	4.84	2
5	4	2				4.24	12.52	4.24	2

In columns L & M we have calculated the distance of each point to the two cluster centers. The cluster center in column O is updated by looking at the cluster that has minimum distance with the data point.

Noting that there is a change in the cluster for a data point, we proceed with the preceding step again, but now with the updated centers. The overall calculation now looks like the following for the two iterations we did:

X	Y	Cluster	Centroid	1	2	Dist-sq	1	2	Cluster
5	0	1	X	2.6	3.2	7.72	7.72	12.24	1
5	2	2	Y	1.4	3	4.24	6.12	4.24	2
3	1	1				0.32	0.32	4.04	1
0	4	2				11.24	13.52	11.24	2
2	1	1				0.52	0.52	5.44	1
4	2	2				1.64	2.32	1.64	2
2	2	1				0.72	0.72	2.44	1
2	3	2				1.44	2.92	1.44	2
1	3	1				4.84	5.12	4.84	2
5	4	2				4.24	12.52	4.24	2

X	Y	Cluster	Centroid	1	2	Dist-sq	1	2	Cluster
5	0	1	X	3	2.833333	5	5	13.69444	1
5	2	2	Y	1	3	5	5	5.694444	1
3	1	1				0	0	4.027778	1
0	4	2				9.027778	18	9.027778	2
2	1	1				1	1	4.694444	1
4	2	2				2	2	2.361111	1
2	2	1				1.694444	2	1.694444	2
2	3	2				0.694444	5	0.694444	2
1	3	2				3.361111	8	3.361111	2
5	4	2				5.694444	13	5.694444	2

We keep on iterating the process until there is no further change in the clusters that data points belong to. If a data point cluster keeps on changing, we would potentially stop after a few iterations.

Properties of the K-means Clustering Algorithm

As noted earlier, the objective of a clustering exercise is to create distinct groups in such a way that the following are true:

1. All points belonging to the same group are as close to each other as possible

2. Each group's center is as far away from other group's center as possible

There are measures that help in assessing the quality of a clustering output based on these objectives.

Let's cement our understanding of the properties through a sample dataset (available in github as "clustering output interpretation.xlsx"). Let's say we have a dataset as follows: two independent variables (*x* and *y*) and the corresponding cluster they belong to (a total of four clusters, just for this example):

x	y	Cluster
0.12	0.40	1
0.04	0.90	1
0.50	0.88	4
0.58	0.30	3
0.84	0.13	2
0.65	0.27	3
0.94	0.01	2
0.51	0.82	4
0.08	0.17	1
0.99	0.84	4

The four cluster centers are as follows:

Cluster	x	y
1	0.080793	0.490584
2	0.891916	0.06916
3	0.616464	0.281198
4	0.670311	0.846514

We calculate the distance of every point with respect to its corresponding cluster center as follows:

	A	B	C	D	E	F	G	H	I	J	K
1	x		y	Cluster	kx	ky		withinss			
2		0.12	0.40	1	0.08	0.49	0.01		Cluster	x	y
3		0.04	0.90	1	0.08	0.49	0.17		1	0.080793	0.490584
4		0.50	0.88	4	0.67	0.85	0.03		2	0.891916	0.0691597
5		0.58	0.30	3	0.62	0.28	0.00		3	0.616464	0.2811979
6		0.84	0.13	2	0.89	0.07	0.01		4	0.670311	0.8465136
7		0.65	0.27	3	0.62	0.28	0.00				
8		0.94	0.01	2	0.89	0.07	0.01				
9		0.51	0.82	4	0.67	0.85	0.03				
10		0.08	0.17	1	0.08	0.49	0.10				
11		0.99	0.84	4	0.67	0.85	0.11				

Note that the column withinss is calculating the distance of each point to its corresponding cluster center. Let's look at the formulas used to arrive at the preceding results:

	A	B	C	D	E	F	G
1		x	y	Cluster	kx	ky	withinss
2		0.11	0.40	1	=VLOOKUP(D2,$I$3:$J$6,2,0)	=VLOOKUP(D2,$I$3:$K$6,3,0)	=((B2-E2)^2+(C2-F2)^2)
3		0.03	0.89	1	=VLOOKUP(D3,$I$3:$J$6,2,0)	=VLOOKUP(D3,$I$3:$K$6,3,0)	=((B3-E3)^2+(C3-F3)^2)
4		0.50	0.87	4	=VLOOKUP(D4,$I$3:$J$6,2,0)	=VLOOKUP(D4,$I$3:$K$6,3,0)	=((B4-E4)^2+(C4-F4)^2)
5		0.57	0.29	3	=VLOOKUP(D5,$I$3:$J$6,2,0)	=VLOOKUP(D5,$I$3:$K$6,3,0)	=((B5-E5)^2+(C5-F5)^2)
6		0.83	0.13	2	=VLOOKUP(D6,$I$3:$J$6,2,0)	=VLOOKUP(D6,$I$3:$K$6,3,0)	=((B6-E6)^2+(C6-F6)^2)
7		0.65	0.26	3	=VLOOKUP(D7,$I$3:$J$6,2,0)	=VLOOKUP(D7,$I$3:$K$6,3,0)	=((B7-E7)^2+(C7-F7)^2)
8		0.94	0.00	2	=VLOOKUP(D8,$I$3:$J$6,2,0)	=VLOOKUP(D8,$I$3:$K$6,3,0)	=((B8-E8)^2+(C8-F8)^2)
9		0.51	0.81	4	=VLOOKUP(D9,$I$3:$J$6,2,0)	=VLOOKUP(D9,$I$3:$K$6,3,0)	=((B9-E9)^2+(C9-F9)^2)
10		0.08	0.17	1	=VLOOKUP(D10,$I$3:$J$6,2,0)	=VLOOKUP(D10,$I$3:$K$6,3,0)	=((B10-E10)^2+(C10-F10)^2)
11		0.99	0.84	4	=VLOOKUP(D11,$I$3:$J$6,2,0)	=VLOOKUP(D11,$I$3:$K$6,3,0)	=((B11-E11)^2+(C11-F11)^2)

	H	I	J	K
1				
2		Cluster	x	y
3		1	=AVERAGEIF($D$2:$D$11,$I3,B$2:B$11)	=AVERAGEIF($D$2:$D$11,$I3,C$2:C$11)
4		2	=AVERAGEIF($D$2:$D$11,$I4,B$2:B$11)	=AVERAGEIF($D$2:$D$11,$I4,C$2:C$11)
5		3	=AVERAGEIF($D$2:$D$11,$I5,B$2:B$11)	=AVERAGEIF($D$2:$D$11,$I5,C$2:C$11)
6		4	=AVERAGEIF($D$2:$D$11,$I6,B$2:B$11)	=AVERAGEIF($D$2:$D$11,$I6,C$2:C$11)

Totss (Total Sum of Squares)

In a scenario where the original dataset itself is considered as a cluster, the midpoint of
the original dataset is considered as the center. *Totss* is the sum of the squared distance
of all points to the dataset center.

Let's look at the formulas:

	A	B	C
1		x	y
2		0.11835658	0.40144412
3		0.03910006	0.895646297
4		0.50450503	0.876858412
5		0.57848256	0.296484798
6		0.83930391	0.13259191
7		0.65444498	0.265910908
8		0.94452834	0.005727543
9		0.51165763	0.819601821
10		0.08492106	0.174661679
11		0.99477051	0.843080552
12			
13	Overall center	=AVERAGE(B2:B11)	=AVERAGE(C2:C11)
14			
15		=(B2-B$13)^2	=(C2-C$13)^2
16		=(B3-B$13)^2	=(C3-C$13)^2
17		=(B4-B$13)^2	=(C4-C$13)^2
18		=(B5-B$13)^2	=(C5-C$13)^2
19		=(B6-B$13)^2	=(C6-C$13)^2
20		=(B7-B$13)^2	=(C7-C$13)^2
21		=(B8-B$13)^2	=(C8-C$13)^2
22		=(B9-B$13)^2	=(C9-C$13)^2
23		=(B10-B$13)^2	=(C10-C$13)^2
24		=(B11-B$13)^2	=(C11-C$13)^2

Total sum of squares would be the sum of all the values from cell B15 to cell C24.

273

Cluster Centers

The *cluster center* of each cluster will be the midpoint (mean) of all the points that fall in the same cluster. For example, in the Excel sheet cluster centers are calculated in columns I and J. Note that this is just the average of all points that fall in the same cluster.

Tot.withinss

Tot.withinss is the sum of the squared distance of all points to their corresponding cluster center .

Betweenss

Betweenss is the difference between totss and tot.withinss.

Implementing K-means Clustering in R

K-means clustering in R is implemented by using the kmeans function, as follows (available as "clustering_code.R" in github):

```
# Lets generate dataset randomly
x=runif(1000)
y=runif(1000)
```

```
data=cbind(x,y)
# One would have to specify the dataset along with the number of clusters
in input
km=kmeans(data,2)
```

The output of km is the major metrics discussed earlier:

```
> str(km)
List of 9
 $ cluster      : int [1:1000] 1 2 2 1 2 1 2 2 1 1 ...
 $ centers      : num [1:2, 1:2] 0.75 0.252 0.482 0.519
  ..- attr(*, "dimnames")=List of 2
  .. ..$ : chr [1:2] "1" "2"
  .. ..$ : chr [1:2] "x" "y"
 $ totss        : num 166
 $ withinss     : num [1:2] 51.1 52.1
 $ tot.withinss: num 103
 $ betweenss    : num 62.4
 $ size         : int [1:2] 501 499
 $ iter         : int 1
 $ ifault       : int 0
 - attr(*, "class")= chr "kmeans"
```

Implementing K-means Clustering in Python

K-means clustering in Python is implemented by functions available in the scikit-learn library as follows (available as "clustering.ipynb" in github):

```
# import packages and dataset
import pandas as pd

import numpy as np

data2=pd.read_csv('D:/data.csv')


# fit k-means with 2 clusters
from sklearn.cluster import KMeans

kmeans = KMeans(n_clusters=2)
kmeans.fit(data2)
```

In that code snippet, we are extracting two clusters from the original dataset named data2. The resulting labels for each data point of which cluster they belong to can be extracted by specifying kmeans.labels_.

Significance of the Major Metrics

As discussed earlier, the objective of the exercise of clustering is to get all the data points that are very close to each other into one group and have groups that are as far away from each other as possible.

Here's another way to say that:

1. Minimize intracluster distance

2. Maximize intercluster distance

Let's look at how the metrics discussed earlier help in achieving the objective. When there is no clustering (that is, all the dataset is considered as one cluster), the overall distance of each point to the cluster center (where there is one cluster) is totss. The moment we introduce clustering in the dataset, the sum of distance of each point within a cluster to the corresponding cluster centre is tot.withinss. Note that as the number of clusters increases, tot.withinss keeps on decreasing.

Consider a scenario where the number of clusters is equal to the number of data points. Tot.withinss is equal to 0 in that scenario, because the distance of each point to the cluster center (which is the point itself) is 0.

Thus, tot.withinss is a measure of intracluster distance. The lower the ratio of tot. withinss / totss, the higher quality is the clustering process.

However, we also need to note that the scenario where tot.withinss = 0 is the scenario where the clustering becomes useless, because each point is a cluster in itself.

In the next section, we'll make use of the metric tot.withinss / totss in a slightly different way.

Identifying the Optimal K

One major question we have not answered yet is how to obtain the optimal k value? In other words, what is the optimal number of clusters within a dataset?

To answer this question, we'll use the metric we used in the last section: the ratio of tot.withinss / totss. To see how the metric varies as we vary the number of clusters (k), see the following code:

```
value_k=c()
value_metric=c()

x=runif(10000)
y=runif(10000)

data=cbind(x,y)
for(i in 1:100){
  km=kmeans(data,i)
  value_k=c(value_k,i)
  metric=km$tot.withinss/km$totss
  value_metric=c(value_metric,metric)
}

plot(value_k,value_metric)
```

We are creating a dataset data with 10,000 randomly initialized x and y values.

Now we'll explore the value of the metric as we vary the k value. The plot looks like Figure 11-8.

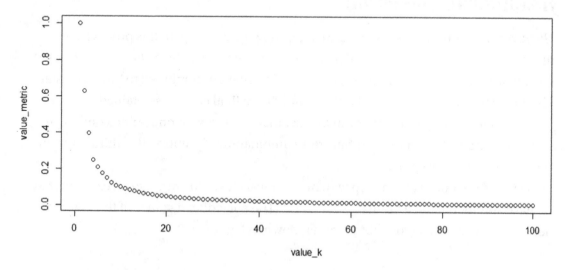

Figure 11-8. *Variation in tot.withinss/totss over different values of k*

Note that as the value of k is increased from $k = 1$ to $k = 2$, there is a steep decrease in the metric, and similarly, the metric decreased when k is reduced from 2 to 4.

However, as k is further reduced, the value of the metric does not decrease by a lot. So, it is prudent to keep the value of k close to 7 because maximum decrease is obtained up to that point, and any further decrease in the metric (tot.withinss / totss) does not correlate well with an increase in k.

Given that the curve looks like an elbow, it sometimes called the *elbow curve*.

Top-Down Vs. Bottom-Up Clustering

So far, in the process of k-means clustering, we don't know the optimal number of clusters, so we keep trying various scenarios with multiple values of k. This is one of the relatively small problems with the bottom-up approach, where we start with an assumption that there is no cluster and slowly keep building multiple clusters, one at a time, till we find the optimal k based on the elbow curve.

Top-down clustering takes an alternative look at the same process. It assumes that each point is a cluster in itself and tries to combine points based on their distance from other points.

Hierarchical Clustering

Hierarchical clustering is a classic form of top-down clustering. In this process, the distance of each point to the rest of points is calculated. Once the distance is calculated, the points that are closest to the point in consideration are combined to form a cluster. This process is repeated across all points, and thus a final cluster is obtained.

The *hierarchical* part comes from the fact that we start with one point, combine it with another point, and then combine this combination of points with a third point, and keep on repeating this process.

Let's have a look at coming up with hierarchical clustering through an example. Say we have six different data points - A,B,C,D,E,F. The eucledian distance of the different data points with respect to other points is shown in Figure 11-9.

Dist	A	B	C	D	E	F
A	0	0.71	5.66	3.61	4.24	3.2
B	0.71	0	4.95	2.92	3.54	2.5
C	5.66	4.95	0	2.24	1.41	2.5
D	3.61	2.92	2.24	0	1	0.5
E	4.24	3.54	1.41	1	0	1.12
F	3.2	2.5	2.5	0.5	1.12	0

Figure 11-9. *Distance of data points*

We see that the minimum distance is between D and F. Thus, we combine D and F. The resulting matrix now looks like Figure 11-10.

Dist	A	B	C	D,F	E
A	0	0.71	5.66	?	4.24
B	0.71	0	4.95	?	3.54
C	5.66	4.95	0	?	1.41
D,F	?	?	?	0	?
E	4.24	3.54	1.41	?	0

Figure 11-10. *The resulting matrix*

How do we fill in the missing values in Figure 11-10? See the following equation:

$$d_{(D,F)\to A} = \min\left(d_{DA}, d_{FA}\right) = \min\left(3.61, 3.20\right) = 3.20$$

Note that, based on the preceding calculation, we replace the missing value in the distance between {D,F} and A with minimum of the distances between DA and FA. Similarly, we would impute with other missing values. We keep proceeding like that until we are left with Figure 11-11.

Dist	(A,B)	((D,F),E),C
(A,B)	0	2.5
((D,F),E),C	2.5	0

Figure 11-11. *The final matrix*

The resulting cluster can now be represented as Figure 11-12.

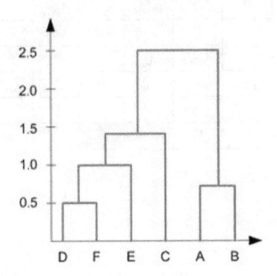

Figure 11-12. *Representing the cluster*

Major Drawback of Hierarchical Clustering

One of the major drawbacks of hierarchical clustering is the large number of calculations one needs to perform.

If there are 100 points in a dataset, say, then the first step is identifying the point that is closest to point 1 and so on for 99 computations. And for the second step, we need to compare the second point's distance with the rest of the 98 points. This makes for a total of $99 \times 100 / 2$, or $n \times (n - 1) / 2$ calculations when there are n data points, only to identify the combination of data points that have the least distance between them among all the combination of data points.

The overall computation becomes extremely complex as the number of data points increase from 100 to 1,000,000. Hence, hierarchical clustering is suited only for small datasets.

Industry Use-Case of K-means Clustering

We've calculated the optimal value of k using the elbow curve of the tot.withinss / totss metric. Let's use a similar calculation for a typical application in building a model.

Let's say we are fitting a model to predict whether a transaction is fraudulent or not using logistic regression. Given that we would be working on all the data points together, this translates into a clustering exercise where $k = 1$ on the overall dataset. And let's say it has an accuracy of 90%. Now let's fit the same logistic regression using $k = 2$, where we have a different model for each cluster. We'll measure the accuracy of using two models on the test dataset.

We keep repeating the exercise by increasing the value of k—that is, by increasing the number of clusters. Optimal k is where we have k different models, one each for each cluster and also the ones that achieve the highest accuracy on top of the test dataset. Similarly, we would use clustering to understand the various segments that are present in the dataset.

Summary

In this chapter, you learned the following:

- K-means clustering helps in grouping data points that are more similar to each other and in forming groups in such a way that the groups are more dissimilar to each other.

- Clustering can form a crucial input in segmentation, operations research, and mathematical modeling.

- Hierarchical clustering takes the opposite approach of k-means clustering in forming the cluster.

- Hierarchical clustering is more computationally intensive to generate when the number of data points is large.

## CHAPTER 12

Principal Component Analysis

Regression typically works best when the ratio of number of data points to number of variables is high. However, in some scenarios, such as clinical trials, the number of data points is limited (given the difficulty in collecting samples from many individuals), and the amount of information collected is high (think of how much information labs give us based on small samples of collected blood).

In these cases, where the ratio of data points to variables is low, one faces difficulty in using the traditional techniques for the following reasons:

- There is a high chance that a majority of the variables are correlated to each other.

- The time taken to run a regression could be very extensive because the number of weights that need to be predicted is large.

Techniques like *principal component analysis* (*PCA*) come to the rescue in such cases. PCA is an unsupervised learning technique that helps in grouping multiple variables into fewer variables without losing much information from the original set of variables.

In this chapter, we will look at how PCA works and get to know the benefits of performing PCA. We will also implement it in Python and R.

Intuition of PCA

PCA is a way to reconstruct the original dataset by using fewer features or variables than the original one had. To see how that might work, consider the following example:

© V Kishore Ayyadevara 2018
V. K. Ayyadevara, *Pro Machine Learning Algorithms*, https://doi.org/10.1007/978-1-4842-3564-5_12

Dep Var	Var 1	Var 2
0	1	10
0	2	20
0	3	30
0	4	40
0	5	50
1	6	60
1	7	70
1	8	80
1	9	90
1	10	100

We'll assume both *Var 1* and *Var 2* are the independent variables used to predict the dependent variable (*Dep Var*). We can see that *Var 2* is highly correlated to *Var 1*, where *Var 2* = (10) × *Var 1*.

A plot of their relation can be seen in Figure 12-1.

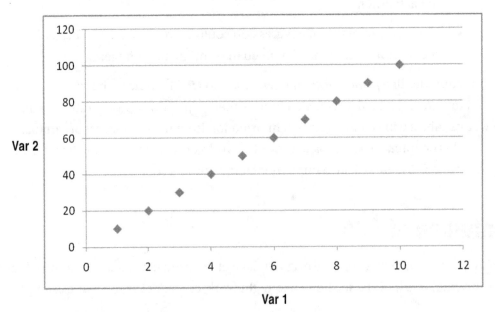

Figure 12-1. *Plotting the relation*

In the figure, we can clearly see that there is a strong relation between the variables. This means the number of independent variables can be reduced.

The equation can be expressed like this:

$$Var2 = 10 \times Var1$$

In other words, instead of using two different independent variables, we could have just used one variable *Var1* and it would have worked out in solving the problem.

Moreover, if we are in a position to view the two variables through a slightly different angle (or, we rotate the dataset), like the one indicated by the arrow in Figure 12-2, we see a lot of variation in horizontal direction and very little in vertical direction.

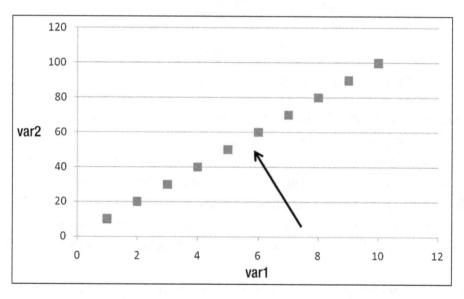

Figure 12-2. *Viewpoint/angle from which data points should be looked at*

Let's complicate our dataset by a bit. Consider a case where the relation between *v1* and *v2* is like that shown in Figure 12-3.

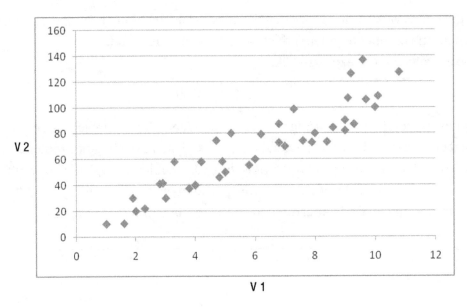

Figure 12-3. *Plotting two variables*

Again, the two variables are highly correlated with each other, though not as perfectly correlated as the previous case.

In such scenario, the first principle component is the line/variable that explains the maximum variance in the dataset and is a linear combination of multiple independent variables. Similarly, the second principal component is the line that is completely uncorrelated (has a correlation of close to 0) to the first principal component and that explains the rest of variance in dataset, while also being a linear combination of multiple independent variables.

Typically the second principal component is a line that is perpendicular to the first principal component (because the next highest variation happens in a direction that is perpendicular to the principal component line).

In general, the nth principal component of a dataset is perpendicular to the $(n - 1)$th principal component of the same dataset.

Working Details of PCA

In order to understand how PCA works, let's look at another example (available as "PCA_2vars.xlsx" in github), where $x1$ and $x2$ are two independent variables that are highly correlated with each other:

x1	x2
1	10.6
2	20.2
3	30.7
4	40.2
5	50.3
6	60.2
7	70.6
8	80.5
9	90.7
10	100.4

Given that a principal component is a linear combination of variables, we'll express it as follows:

$$PC1 = w_1 \times x1 + w_2 \times x2$$

Similarly, the second principal component is perpendicular to the original line, as follows:

$$PC2 = -w_2 \times x1 + w_1 \times x2$$

The weights w_1 and w_2 are randomly initialized and should be iterated further to obtain the optimal ones.

Let's revisit the objective and constraints that we have, while solving for w_1 and w_2:

- *Objective*: Maximize *PC1* variance.

- *Constraint*: Overall variance in principal components should be equal to the overall variance in original dataset (as the data points did not change, but only the angle from which we view the data points changed).

Let's initialize the principal components in the dataset we created earlier:

	A	B	C	D	E	F	G	H
1	x1	x2	pc1	pc2				
2	1	10.6	11.60	9.60			w1	1
3	2	20.2	22.20	18.20			w2	1
4	3	30.7	33.70	27.70				
5	4	40.2	44.20	36.20				
6	5	50.3	55.30	45.30				
7	6	60.2	66.20	54.20				
8	7	70.6	77.60	63.60				
9	8	80.5	88.50	72.50				
10	9	90.7	99.70	81.70				
11	10	100.4	110.40	90.40				

The formulas for *PC1* and *PC2* can be visualized as follows:

	A	B	C	D	E	F	G	H
1	x1	x2	pc1	pc2				
2	1	10.6	=A2*$H$2+B2*$H$3	=-$H$3*A2+$H$2*B2			w1	1
3	2	20.2	=A3*$H$2+B3*$H$3	=-$H$3*A3+$H$2*B3			w2	1
4	3	30.7	=A4*$H$2+B4*$H$3	=-$H$3*A4+$H$2*B4				
5	4	40.2	=A5*$H$2+B5*$H$3	=-$H$3*A5+$H$2*B5				
6	5	50.3	=A6*$H$2+B6*$H$3	=-$H$3*A6+$H$2*B6				
7	6	60.2	=A7*$H$2+B7*$H$3	=-$H$3*A7+$H$2*B7				
8	7	70.6	=A8*$H$2+B8*$H$3	=-$H$3*A8+$H$2*B8				
9	8	80.5	=A9*$H$2+B9*$H$3	=-$H$3*A9+$H$2*B9				
10	9	90.7	=A10*$H$2+B10*$H$3	=-$H$3*A10+$H$2*B				
11	10	100.4	=A11*$H$2+B11*$H$3	=-$H$3*A11+$H$2*B				

Now that we have initialized the principal component variables, we'll bring in the objective and constraints:

	A	B	C	D	E	F	G	H
1	x1	x2	pc1	pc2				
2	1	10.6	11.6	9.6			w1	1
3	2	20.2	22.2	18.2			w2	1
4	3	30.7	33.7	27.7				
5	4	40.2	44.2	36.2			PC variance	927.88
6	5	50.3	55.3	45.3			Original variance	1,855.75
7	6	60.2	66.2	54.2				
8	7	70.6	77.6	63.6			Difference beteen	
9	8	80.5	88.5	72.5			original and PC variance	927.88
10	9	90.7	99.7	81.7				
11	10	100	110.4	90.4			PC1 variance	1,111.41

Note that *PC* variance = *PC1* variance + *PC2* variance.

Original variance = *x1* variance + *x2* variance

We calculate the difference between original and PC variance since our constraint is to maintain the same variance as original dataset in the principal component transformed dataset. Here are their formulas:

	A	B	C	D	E	F	G	H
1	x1	x2	pc1	pc2				
2	1	10.6	=A2*$H$2+B2*$H$3	=-$H$3*A2+$H$2*B2			w1	1
3	2	20.2	=A3*$H$2+B3*$H$3	=-$H$3*A3+$H$2*B3			w2	1
4	3	30.7	=A4*$H$2+B4*$H$3	=-$H$3*A4+$H$2*B4				
5	4	40.2	=A5*$H$2+B5*$H$3	=-$H$3*A5+$H$2*B5			PC variance	=VAR(A2:A11)+VAR(B2:B11)
6	5	50.3	=A6*$H$2+B6*$H$3	=-$H$3*A6+$H$2*B6			Original variance	=VAR(C2:C11)+VAR(D2:D11)
7	6	60.2	=A7*$H$2+B7*$H$3	=-$H$3*A7+$H$2*B7				
8	7	70.6	=A8*$H$2+B8*$H$3	=-$H$3*A8+$H$2*B8			Difference beteen	
9	8	80.5	=A9*$H$2+B9*$H$3	=-$H$3*A9+$H$2*B9			original and PC variance	=ABS(H5-H6)
10	9	90.7	=A10*$H$2+B10*$H$3	=-$H$3*A10+$H$2*B10				
11	10	100.4	=A11*$H$2+B11*$H$3	=-$H$3*A11+$H$2*B11			PC1 variance	=VAR(C2:C11)

Once the dataset is initialized, we will proceed with identifying the optimal values of w_1 and w_2 that satisfy our objective and constraint.

Let us look at how do we achieve that through Excel's Solver add-in:

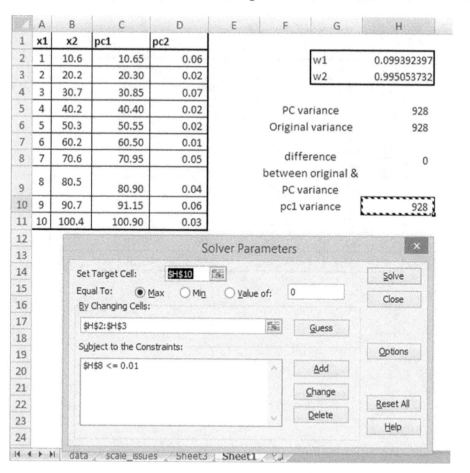

Note that the objective and criterion that we specified earlier are met:

- *PC1* variance is maximized.

- There is hardly any difference between the original dataset variance and the principal component dataset variance. (We have allowed for a small difference of less than 0.01 only so that Excel is able to solve it because there may be some rounding-off errors.)

Note that *PC1* and *PC2* are now highly uncorrelated with each other, and *PC1* explains the highest variance across all variables. Moreover, x2 has a higher weightage in determining *PC1* than *x1* (as is evident from the derived weight values).

In practice, once a principal component is arrived at, it is centered around the corresponding mean value—that is, each value within the principal component column would be subtracted by the average of the original principal component column:

	A	B	C	D	E	F
	x1	x2	pc1	pc2	final pc1	final pc2
1						
2	1	10.6	10.65	0.06	(45.07)	0.02
3	2	20.2	20.30	0.02	(35.41)	(0.02)
4	3	30.7	30.85	0.07	(24.87)	0.03
5	4	40.2	40.40	0.02	(15.31)	(0.02)
6	5	50.3	50.55	0.02	(5.16)	(0.01)
7	6	60.2	60.50	0.01	4.79	(0.02)
8	7	70.6	70.95	0.05	15.23	0.01
9	8	80.5	80.90	0.04	25.18	0.00
10	9	90.7	91.15	0.06	35.43	0.02
11	10	100.4	100.90	0.03	45.18	(0.01)

The formulas used to derive the preceding dataset are shown here:

	A	B	C	D	E	F
1	x1	x2	pc1	pc2	final pc1	final pc2
2	1	10.6	=$J$2*A2+$J$3*B2	=-$J$3*A2+$J$2*B2	=C2-AVERAGE(C$2:C$11)	=D2-AVERAGE(D$2:D$11)
3	2	20.2	=$J$2*A3+$J$3*B3	=-$J$3*A3+$J$2*B3	=C3-AVERAGE(C$2:C$11)	=D3-AVERAGE(D$2:D$11)
4	3	30.7	=$J$2*A4+$J$3*B4	=-$J$3*A4+$J$2*B4	=C4-AVERAGE(C$2:C$11)	=D4-AVERAGE(D$2:D$11)
5	4	40.2	=$J$2*A5+$J$3*B5	=-$J$3*A5+$J$2*B5	=C5-AVERAGE(C$2:C$11)	=D5-AVERAGE(D$2:D$11)
6	5	50.3	=$J$2*A6+$J$3*B6	=-$J$3*A6+$J$2*B6	=C6-AVERAGE(C$2:C$11)	=D6-AVERAGE(D$2:D$11)
7	6	60.2	=$J$2*A7+$J$3*B7	=-$J$3*A7+$J$2*B7	=C7-AVERAGE(C$2:C$11)	=D7-AVERAGE(D$2:D$11)
8	7	70.6	=$J$2*A8+$J$3*B8	=-$J$3*A8+$J$2*B8	=C8-AVERAGE(C$2:C$11)	=D8-AVERAGE(D$2:D$11)
9	8	80.5	=$J$2*A9+$J$3*B9	=-$J$3*A9+$J$2*B9	=C9-AVERAGE(C$2:C$11)	=D9-AVERAGE(D$2:D$11)
10	9	90.7	=$J$2*A10+$J$3*B10	=-$J$3*A10+$J$2*B10	=C10-AVERAGE(C$2:C$11)	=D10-AVERAGE(D$2:D$11)
11	10	100.4	=$J$2*A11+$J$3*B11	=-$J$3*A11+$J$2*B11	=C11-AVERAGE(C$2:C$11)	=D11-AVERAGE(D$2:D$11)

Scaling Data in PCA

One of the major pre-processing steps in PCA is to scale the variables. Consider the following scenario: we are performing PCA on two variables. One variable has a range of values from 0–100, and another variable has a range of values from 0–1.

Given that, using PCA, we are trying to capture as much variation in the dataset as possible, the first principal component will give a very high weightage to the variable that has maximum variance (in our case, *Var1*) when compared to the variable with low variance.

Hence, when we work out w_1 and w_2 for the principal component, we will end up with a w_1 that is close to 0 and a w_2 that is close to 1 (where w_2 is the weight in PC1 corresponding to the higher ranged variable). To avoid this, it is advisable to scale each variable so that both of them have similar range, due to which variance can be comparable.

Extending PCA to Multiple Variables

So far, we have seen building a PCA where there are two independent variables. In this section, we will consider how to hand-build a PCA where there are more than two independent variables.

Consider the following dataset (available as "PCA_3vars.xlsx" in github):

x1	x2	x3
1	10.6	112.7
2	20.2	205.4
3	30.7	314.5
4	40.2	412
5	50.3	506.7
6	60.2	602
7	70.6	712.8
8	80.5	813.3
9	90.7	908.1
10	100.4	1011.5

Unlike a two-variable PCA, in a more than 2-dimensional PCA, we'll initialize the weights in a slightly different way. The weights are initialized randomly—but in matrix form, as follows:

	x1	x2	x3
PC1	0.49	0.89	0.92
PC2	1	0.62	0.83
PC3	0.34	0.38	0.94

From this matrix, we can consider $PC1 = 0.49 \times x1 + 0.89 \times x2 + 0.92 \times x3$. $PC2$ and $PC3$ would be worked out similarly. If there were four independent variables, we would have had a 4×4-weight matrix.

Let's look at the objective and constraints we might have:

- *Objective*: Maximize $PC1$ variance.

- *Constraints*: Overall PC variance should be equal to overall original dataset variance. $PC1$ variance should be greater than $PC2$ variance, $PC1$ variance should be greater than $PC3$ variance, and $PC2$ variance should be greater than $PC3$ variance.

	A	B	C	D	E	F	G	H	I	J (x1)	K (x2)	L (x3)
1	x1	x2	x3	PC1	PC2	PC3				x1	x2	x3
2	1	10.6	112.7	41.998	87.081	115.031			PC1	0.74	0.49	0.32
3	2	20.2	205.4	77.106	159.322	210.402			PC2	0.28	0.64	0.71
4	3	30.7	314.5	117.903	243.783	321.947			PC3	0.97	0.66	0.95
5	4	40.2	412	154.498	319.368	421.812						
6	5	50.3	506.7	190.491	393.349	519.413			PC variance	164,789.62		
7	6	60.2	602	226.578	467.628	617.452			Original variance	92,609.22		
8	7	70.6	712.8	267.87	553.232	730.546						
9	8	80.5	813.3	305.621	631.203	833.525			PC1 variance	12,992.38	113.98	
10	9	90.7	908.1	341.695	705.319	931.287			PC2 variance	55,331.07	235.23	
11	10	100.4	1011.5	380.276	785.221	1036.889			PC3 variance	96,466.18	310.59	
12												
13							Diff between original and pc variance			72,180.40		

Solver Parameters

Set Target Cell: $J$9

Equal To: ● Max ○ Min ○ Value of: 0

By Changing Cells:
$J$2:$L$4

Subject to the Constraints:
$J$10 <= $J$9
$J$11 <= $J$10
$J$13 <= 3

Buttons: Solve, Close, Guess, Options, Add, Change, Reset All, Delete, Help

Sheets: pca_3vars Sheet2 Sheet3

Solving for the preceding would result in the optimal weight combination that satisfies our criterion. Note that the output from Excel could be slightly different from the output you would see in Python or R, but the output of Python or R is likely to have higher PC1 variance when compared to the output of Excel, due to the underlying algorithm used in solving. Also note that, even though ideally we would have wanted the difference between original and PC variance to be 0, for practical reasons of executing the optimization using Excel solver we have allowed the difference to be a maximum of 3.

Similar to the scenario of the PCA with two independent variables, it is a good idea to scale the inputs before processing PCA. Also, note that *PC1* explains the highest variation after solving for the weights, and hence *PC2* and *PC3* can be eliminated because they explain very little of the original dataset variance.

CHOOSING THE NUMBER OF PRINCIPAL COMPONENTS TO CONSIDER

There is no single prescribed method to choosing the number of principal components. In practice, a rule of thumb is to choose the minimum number of principal components that cumulatively explain 80% of the total variance in the dataset.

Implementing PCA in R

PCA can be implemented in R using the built-in function prcomp. Have a look at the following implementation (available as "PCA R.R" in github):

```
t=read.csv('D:/Pro ML book/PCA/pca_3vars.csv')
pca=prcomp(t)
pca
```

The output of pca is as follows:

```
> pca
Standard deviations:
[1] 304.31745149    0.32640440    0.01998034

Rotation:
            PC1         PC2          PC3
x1 0.009948083   0.1092539 -0.99396410
x2 0.099595009   0.9889624  0.10970088
x3 0.994978326  -0.1000852 -0.00104286
```

The standard deviation values here are the same as the standard deviation values of PC variables. The rotation values are the same as the weight values that we initialized earlier.

A more detailed version of the outputs can be obtained by using str(pca), the output of which looks like the following:

```
> str(pca)
List of 5
 $ sdev     : num [1:3] 304.317 0.326 0.02
 $ rotation: num [1:3, 1:3] 0.00995 0.0996 0.99498 0.10925 0.98896 ...
 ..- attr(*, "dimnames")=List of 2
 .. ..$ : chr [1:3] "x1" "x2" "x3"
 .. ..$ : chr [1:3] "PC1" "PC2" "PC3"
 $ center   : Named num [1:3] 5.5 55.4 559.9
 ..- attr(*, "names")= chr [1:3] "x1" "x2" "x3"
 $ scale    : logi FALSE
 $ x        : num [1:10, 1:3] -449.5 -356.3 -246.7 -148.7 -53.4 ...
 ..- attr(*, "dimnames")=List of 2
 .. ..$ : NULL
 .. ..$ : chr [1:3] "PC1" "PC2" "PC3"
 - attr(*, "class")= chr "prcomp"
```

From this, we notice that apart from the standard deviation of PC variables and the weight matrix, pca also provides the transformed dataset.

We can access the transformed dataset by specifying pca$x.

Implementing PCA in Python

Implementing PCA in Python is done using the scikit learn library, as follows (available as "PCA.ipynb" in github):

```
# import packages and dataset
import pandas as pd
import numpy as np
from sklearn.decomposition import PCA
data=pd.read_csv('F:/course/pca/pca.csv')
from sklearn.decomposition import PCA
pca = PCA(n_components=2)
pca.fit(data)
```

We see that we fit as many components as the number of independent variables and fit PCA on top of the data.

Once the data is fit, transform the original data into transformed data, as follows:

```
x_pca = pca.transform(data)
```

```
pca.components_
array([[ 0.08102074,  0.99671242],
       [-0.99671242,  0.08102074]])
```

components_ is the same as weights associated with the principal components. x_pca is the transformed dataset.

```
print(pca.explained_variance_ratio_)
```

```
[0.9988235 0.0011765]
```

explained_variance_ratio_ provides the amount of variance explained by each principal component. This is very similar to the standard deviation output in R, where R gives us the standard deviation of each principal component. PCA in Python's scikit learn transformed it slightly and gave us the amount of variance out of the original variance explained by each variable.

Applying PCA to MNIST

MNIST is a handwritten digit recognition task. A 28 × 28 image is unrolled, where each pixel value is represented in a column. Based on that, one is expected to predict if the output is one of the numbers between 0 and 9.

Given that there are a total of 784 columns, intuitively we should observe one of the following:

- Columns with zero variance

- Columns with very little variance

- Columns with high variance

In a way, PCA helps us in eliminating low- and no-variance columns as much as possible while still achieving decent accuracy with a limited number of columns.

Let's see how to achieve that reduction in number of columns without losing out on much variance through the following example (available as "PCA mnist.R" in github):

```
# Load dataset
t=read.csv("D:/Pro ML book/PCA/train.csv")
# Keep the independent variables only, as PCA is on indep. vars
t$Label = NULL
# scale dataset by 255, as it is the mximum possible value in pixels
t=t/255
# Apply PCA
pca=prcomp(t)
str(pca)
# Check the variance explained
cumsum((pca$sdev)^2)/sum(pca$sdev^2)
```

The output of the preceding code is as follows:

```
> cumsum((pca$sdev)^2)/sum(pca$sdev^2)
 [1]  0.0974892 0.1690923 0.2305512 0.2843441 0.3332869 0.3763189
 [7]  0.4090894 0.4380103 0.4656793 0.4891680 0.5101612 0.5307512
[13]  0.5477767 0.5647045 0.5805158 0.5953482 0.6085450 0.6213723
[19]  0.6332520 0.6447796 0.6555015 0.6656534 0.6753025 0.6844309
[25]  0.6933073 0.7016949 0.7098135 0.7175875 0.7249939 0.7318605
[31]  0.7384403 0.7448283 0.7508219 0.7567110 0.7623544 0.7677640
[37]  0.7728563 0.7777313 0.7824870 0.7871524 0.7916819 0.7961318
[43]  0.8003144 0.8042894 0.8081348 0.8118840 0.8154941 0.8189794
[49]  0.8223442 0.8255516 0.8287063 0.8317977 0.8347348 0.8376002
[55]  0.8404078 0.8431040 0.8457623 0.8483253 0.8508635 0.8533253
[61]  0.8557224 0.8581098 0.8603857 0.8626009 0.8647402 0.8668016
[67]  0.8688301 0.8707898 0.8727262 0.8746111 0.8764786 0.8782953
[73]  0.8800642 0.8817901 0.8834513 0.8850844 0.8866904 0.8882351
[79]  0.8897036 0.8911274 0.8925384 0.8939407 0.8953290 0.8966832
[85]  0.8980062 0.8993140 0.9006108 0.9018532 0.9030757 0.9042719
```

From this we can see that the first 43 principal components explain ~80% of the total variance in the original dataset. Instead of running a model on all 784 columns, we could run the model on the first 43 principal components without losing much information and hence without losing out much on accuracy.

Summary

- PCA is a way of reducing the number of independent variables in a dataset and is particularly applicable when the ratio of data points to independent variables is low.

- It is a good idea to scale independent variables before applying PCA.

- PCA transforms a linear combination of variables such that the resulting variable expresses the maximum variance within the combination of variables.

CHAPTER 13

Recommender Systems

We see recommendations everywhere. Recommender systems are aimed at

- Minimizing the effort of users to search for a product

- Reminding users about sessions they closed earlier

- Helping users discover more products

For example, here are common instances of recommender systems:

- Recommender widgets in e-commerce websites

- Recommended items sent to email addresses

- Recommendations of friends/contacts in social network sites

Imagine a scenario in which e-commerce customers are not given product recommendations. The customers would not be able to do the following:

- Identify similar products to the products they are viewing

- Know whether the product is fairly priced

- Find accessories or complementary products

That's why recommender systems often boost sales by a considerable amount. In this chapter, you will learn the following:

- To predict the rating a user would give (or likelihood a user would purchase) an item using

  - Collaborative filtering

  - Matrix factorization

© V Kishore Ayyadevara 2018
V. K. Ayyadevara, *Pro Machine Learning Algorithms*, https://doi.org/10.1007/978-1-4842-3564-5_13

- Euclidian and cosine similarity measures
- How to implement recommendation algorithms in Excel, Python, and R

A recommender system is almost like a friend. It infers your preferences and provides you with options that are personalized to you. There are multiple ways of building a recommender system, but the goal is for it to be a way of relating the user to a set of other users, a way of relating an item to a set of other items, or a combination of both.

Given that recommending is about relating one user/item to another, it translates to a problem of *k-nearest neighbors*: identifying the few that are very similar and then basing a prediction based on the preferences exhibited by the majority of nearest neighbors.

Understanding k-nearest Neighbors

A *nearest neighbor* is the entity (a data point, in the case of a dataset) that is closest to the entity under consideration. Two entities are *close* if the distance between them is small.

Consider three users with the following attributes:

User	Weight
A	60
B	62
C	90

We can intuitively conclude that users A and B are more similar to each other in terms of weight when compared to C.

Let's add one more attribute of users—age:

User	Weight	Age
A	60	30
B	62	35
C	90	30

The "distance" between user A and B can be measured as:

$$\sqrt{((62-60)^2+(35-30)^2)}$$

This kind of distance between users is calculated similarly to the way the distance between two points is calculated.

However, you need to be a little careful when calculating distance using multiple variables. The following example can highlight the pitfalls of distance calculation:

Car model	Max speed attainable	No. of gears
A	100	4
B	110	5
C	100	5

In the preceding table, if we were measuring to see the similarity between cars using the traditional "distance" metric, we might conclude that models A and C are most similar to each other (even though their number of gears are different). Yet intuitively we know that B and C are more similar to each other than A and C, because they have identical number of gears and their max attainable speeds are similar.

The discrepancy highlights the issue of *scale of variables*, where one variable has a very high magnitude when compared to the other variable. To get around this issue, typically we would normalize the variables before proceeding further with distance calculations. *Normalizing* variables is a process of bringing all the variables to a uniform scale.

There are multiple ways normalizing a variable:

- Divide each variable by the maximum value of the variable (bringing all the values between –1 and 1)

- Find the Z-score of each data point of the variable. *Z-score* is (value of data point – mean of variable) / (standard deviation of the variable).

- Divide each variable by the (maximum – minimum) value of the variable (called *min max scaling*).

Steps like these help normalize variables, thereby preventing the issues with scaling.

Once the distance of a data point to other data points is obtained—in the case of recommender systems, that is, once the nearest items to a given item are identified—the system will recommend these items to a user if it learns that the user has historically liked a majority of the nearest neighborhood items.

The *k* in k-nearest neighbors stands for the number of nearest neighbors to consider while taking a majority vote on whether the user likes the nearest neighbors or not. For example, if the user likes 9 out of 10 (k) nearest neighbors to an item, we'll recommend the item to user. Similarly, if the user likes only 1 out of 10 nearest neighbors of the item, we'll not recommend the item to user (because the liked items are in minority).

Neighborhood-based analysis takes into account the way in which multiple users can collaboratively help predict whether a user might like something or not.

With this background, we'll move on and look at the evolution of recommender system algorithms.

Working Details of User-Based Collaborative Filtering

User-based refers, of course, to something based on users. *Collaborative* means using some relation (similarity) between users. And *filtering* refers to filtering out some users from all users.

To get a sense of user-based collaborative filtering (UBCF), consider the following example (available as "ubcf.xlsx" in github):

User/ Movie	Just My Luck	Lady in the Water	Snakes on a Plane	Superman Returns	The Night Listener	You Me and Dupree
Claudia Puig	3		3.5	4	4.5	2.5
Gene Seymour	1.5	3	3.5	5	3	3.5
Jack Matthews		3	4	5	3	3.5
Lisa Rose	3	2.5	3.5	3.5	3	2.5
Mick LaSalle	2	3	4	3	3	2
Toby			4.5	4		1

Let's say we are interested in knowing the rating that user Claudia Puig would give to the movie *Lady in the Water*. We'll begin by finding out the most similar user to Claudia. User similarity can be calculated in several ways. Here are two of the most common ways of calculating similarity:

- Euclidian distance between users

- Cosine similarity between users

Euclidian Distance

Calculating the Euclidian distance of Claudia with every other user can be done as follows (available in "Eucledian distance" sheet of "ubcf.xlsx" file in github):

	C	D	E	F
1				
2				
3	User/ Movie	Just My Luck	Lady in the Water	Snakes on a Plane
4	Claudia Puig	3		3.5
5	Gene Seymour	1.5	3	3.5
6	Jack Matthews		3	4
7	Lisa Rose	3	2.5	3.5
8	Mick LaSalle	2	3	4
9	Toby			4.5
10				
11				
12	Gene Seymour	=IF(OR(D$4="",D5=""),"",(D$4-D5)^2)	=IF(OR(E$4="",E5=""),"",(E$4-E5)^2)	=IF(OR(F$4="",F5=""),"",(F$4-F5)^2)
13	Jack Matthews	=IF(OR(D$4="",D6=""),"",(D$4-D6)^2)	=IF(OR(E$4="",E6=""),"",(E$4-E6)^2)	=IF(OR(F$4="",F6=""),"",(F$4-F6)^2)
14	Lisa Rose	=IF(OR(D$4="",D7=""),"",(D$4-D7)^2)	=IF(OR(E$4="",E7=""),"",(E$4-E7)^2)	=IF(OR(F$4="",F7=""),"",(F$4-F7)^2)
15	Mick LaSalle	=IF(OR(D$4="",D8=""),"",(D$4-D8)^2)	=IF(OR(E$4="",E8=""),"",(E$4-E8)^2)	=IF(OR(F$4="",F8=""),"",(F$4-F8)^2)
16	Toby	=IF(OR(D$4="",D9=""),"",(D$4-D9)^2)	=IF(OR(E$4="",E9=""),"",(E$4-E9)^2)	=IF(OR(F$4="",F9=""),"",(F$4-F9)^2)

We aren't seeing the complete picture due to space and formatting constraints, but essentially the same formula is applied across columns.

The distance of every other user to Claudia for each movie is as follows:

	C	D	E	F	G	H	I	J
1								
2								
3	User/ Movie	Just My Luck	Lady in the Water	Snakes on a Plane	Superman Returns	The Night Listener	You Me and Dupree	
4	Claudia Puig	3		3.5	4	4.5	2.5	
5	Gene Seymour	1.5	3	3.5	5	3	3.5	
6	Jack Matthews		3	4	5	3	3.5	
7	Lisa Rose	3	2.5	3.5	3.5	3	2.5	
8	Mick LaSalle	2	3	4	3	3	2	
9	Toby			4.5	4		1	
10								
11	Distance							Overall distance
12	Gene Seymour	2.25		0	1	2.25	1	1.30
13	Jack Matthews			0.25	1	2.25	1	1.13
14	Lisa Rose	0		0	0.25	2.25	0	0.50
15	Mick LaSalle	1		0.25	1	2.25	0.25	0.95
16	Toby			1	0		2.25	1.08

Note that the *overall distance* value is the average of all the distances where both users have rated a given movie. Given that Lisa Rose is the user who has the least overall distance with Claudia, we will consider the rating provided by Lisa as the rating that Claudia is likely to give to the movie *Lady in the Water*.

One major issue to be considered in such a calculation is that some users may be soft critics and some users may be harsher critics. Users A and B may have implicitly had similar experiences watching a given movie, but explicitly their ratings might be different.

Normalizing for a User

Given that users differ in their levels of criticism, we need to make sure that we get around that problem. Normalization can help here.

We could normalize for a user as follows:

1.   Take the average rating across all movies of a given user.

2.   Take the difference between each individual movie and the average rating of the user.

By taking the difference between a rating of an individual movie and the average rating of the user, we would know whether they liked a movie *more* than the average movie they watch, *less* than the average movie they watch, or *equal to* the average movie they watch.

Let's look at how this is done:

User/ Movie	Just My Luck	Lady in the Water	Snakes on a Plane	Superman Returns	The Night Listener	You Me and Dupree	Average rating
Claudia Puig	3		3.5	4	4.5	2.5	3.50
Gene Seymour	1.5	3	3.5	5	3	3.5	3.25
Jack Matthews		3	4	5	3	3.5	3.70
Lisa Rose	3	2.5	3.5	3.5	3	2.5	3.00
Mick LaSalle	2	3	4	3	3	2	2.83
Toby			4.5	4		1	3.17

Claudia Puig	-0.50		0.00	0.50	1.00	-1.00	
Gene Seymour	-1.75	-0.25	0.25	1.75	-0.25	0.25	
Jack Matthews		-0.70	0.30	1.30	-0.70	-0.20	
Lisa Rose	0.00	-0.50	0.50	0.50	0.00	-0.50	
Mick LaSalle	-0.83	0.17	1.17	0.17	0.17	-0.83	
Toby			1.33	0.83		-2.17	

The formulas for the above are as follows (available in "Normalizing user" sheet of "ubcf.xlsx" file in github):

	C	D	E	F	G	H	I	J
3	User/ Movie	Just My Luck	Lady in the Water	Snakes on a Plane	Superman Returns	The Night Listener	You Me and Dupre	Average rating
4	Claudia Puig	3		3.5	4	4.5	2.5	=AVERAGE(D4:I4)
5	Gene Seymour	1.5	3	3.5	5	3	3.5	=AVERAGE(D5:I5)
6	Jack Matthews		3	4	5	3	3.5	=AVERAGE(D6:I6)
7	Lisa Rose	3	2.5	3.5	3.5	3	2.5	=AVERAGE(D7:I7)
8	Mick LaSalle	2	3	4	3	3	2	=AVERAGE(D8:I8)
9	Toby			4.5	4		1	=AVERAGE(D9:I9)
10								
11	Claudia Puig	=IF(D4="","",D4-$J4)	=IF(E4="","",E4-$J4)	=IF(F4="","",F4-$J4)	=IF(G4="","",G4-$J4)	=IF(H4="","",H4-$J4)	=IF(I4="","",I4-$J4)	
12	Gene Seymour	=IF(D5="","",D5-$J5)	=IF(E5="","",E5-$J5)	=IF(F5="","",F5-$J5)	=IF(G5="","",G5-$J5)	=IF(H5="","",H5-$J5)	=IF(I5="","",I5-$J5)	
13	Jack Matthews	=IF(D6="","",D6-$J6)	=IF(E6="","",E6-$J6)	=IF(F6="","",F6-$J6)	=IF(G6="","",G6-$J6)	=IF(H6="","",H6-$J6)	=IF(I6="","",I6-$J6)	
14	Lisa Rose	=IF(D7="","",D7-$J7)	=IF(E7="","",E7-$J7)	=IF(F7="","",F7-$J7)	=IF(G7="","",G7-$J7)	=IF(H7="","",H7-$J7)	=IF(I7="","",I7-$J7)	
15	Mick LaSalle	=IF(D8="","",D8-$J8)	=IF(E8="","",E8-$J8)	=IF(F8="","",F8-$J8)	=IF(G8="","",G8-$J8)	=IF(H8="","",H8-$J8)	=IF(I8="","",I8-$J8)	
16	Toby	=IF(D9="","",D9-$J9)	=IF(E9="","",E9-$J9)	=IF(F9="","",F9-$J9)	=IF(G9="","",G9-$J9)	=IF(H9="","",H9-$J9)	=IF(I9="","",I9-$J9)	

Now that we have normalized for a given user, we calculate which user is most similar to Claudia the same way we calculated user similarity earlier. The only difference is that now we will calculate distance based on normalized ratings, not the original ratings:

	C	D	E	F	G	H	I	J
3	User/ Movie	Just My Luck	Lady in the Water	Snakes on a Plane	Superman Returns	The Night Listener	You Me and Dupree	Average rating
4	Claudia Puig	3		3.5	4	4.5	2.5	3.50
5	Gene Seymour	1.5	3	3.5	5	3	3.5	3.25
6	Jack Matthews		3	4	5	3	3.5	3.70
7	Lisa Rose	3	2.5	3.5	3.5	3	2.5	3.00
8	Mick LaSalle	2	3	4	3	3	2	2.83
9	Toby			4.5	4		1	3.17
10								
11	Claudia Puig	-0.50		0.00	0.50	1.00	-1.00	
12	Gene Seymour	-1.75	-0.25	0.25	1.75	-0.25	0.25	
13	Jack Matthews		-0.70	0.30	1.30	-0.70	-0.20	
14	Lisa Rose	0.00	-0.50	0.50	0.50	0.00	-0.50	
15	Mick LaSalle	-0.83	0.17	1.17	0.17	0.17	-0.83	
16	Toby			1.33	0.83		-2.17	
17								
18								Average distance
19	Gene Seymour	1.56		0.06	1.56	1.56	1.56	1.26
20	Jack Matthews			0.09	0.64	2.89	0.64	1.07
21	Lisa Rose	0.25		0.25	-	1.00	0.25	0.35
22	Mick LaSalle	0.11		1.36	0.11	0.69	0.03	0.46
23	Toby			1.78	0.11		1.36	1.08

We can see that Lisa Rose is still the least distant (or closest, or most similar) user to Claudia Puig. Lisa rated *Lady in the Water* 0.50 units lower than her average movie rating of 3.00, which works out to ~8% lower than her average rating. Given that Lisa is the most similar user to Claudia, we would expect Claudia's rating to likewise be 8% less than *her* average rating, which works out to the following:

$$3.5 \times (1 - 0.5 / 3) = 2.91$$

Issue with Considering a Single User

So far, we have considered the single user who is most similar to Claudia. In practice, more is always better—that is, identifying the weighted average rating that k most similar users to a given user give is better than identifying the rating of the most similar user.

However, we need to note that not all k users are equally similar. Some are more similar, and others are less similar. In other words, some users' ratings should be given more weightage, and other users' ratings should be given less weightage. But using the distance-based metric, there is no easy way to come up with a similarity metric.

Cosine similarity as a metric comes in handy to solve this problem.

Cosine Similarity

We can look at cosine similarity by going through an example. Consider the following matrix:

	Movie1	Movie2	Movie3
User1	1	2	2
User2	2	4	4

In the preceding table, we see that both users' ratings are highly correlated with each other. However, there is a difference in the magnitude of ratings.

If we were to compute Euclidian distance between the two users, we would notice that the two users are very different from each other. But we can see that the two users are similar in the direction (trend) of their ratings, though not in the magnitude of their ratings. The problem where the trend of users is similar but not the magnitude can be solved using cosine similarity between the users.

Cosine similarity between two users is defined as follows:

$$similarity = \cos(\theta) = \frac{A \cdot B}{\|A\|_2 \|B\|_2} = \frac{\sum_{i=1}^{n} A_i B_i}{\sqrt{\sum_{i=1}^{n} A_i^2} \sqrt{\sum_{i=1}^{n} B_i^2}}$$

A and B are the vectors corresponding to user 1 and user 2 respectively. Let's look how similarity is calculated for the preceding matrix:

- Numerator of the given formula = $(1 \times 2 + 2 \times 4 + 2 \times 4) = 18$

- Denominator of the given formula = $\sqrt{(1^2 + 2^2 + 2^2)} \times \sqrt{(2^2 + 4^2 + 4^2)} = \sqrt{(9)} \times \sqrt{(36)} = 3 \times 6 = 18$

- Similarity = $18 / 18 = 1$.

Based on the given formula, we can see that, on the basis of cosine similarity, we are in a position to assign high similarity to users that are directionally correlated but not necessarily in magnitude.

Cosine similarity on the rating matrix that we originally calculated (in the Euclidian distance calculation) earlier would be calculated in a similar way to how we calculated the preceding formula. The steps for cosine similarity calculation remain the same:

1. Normalize users.

2. Calculate the cosine similarity of rest of the users for a given user.

To illustrate how we calculate cosine similarity, let's calculate the similarity of Claudia with every other user (available in "cosine similarity" sheet of "ubcf.xlsx" file in github):

1. Normalize user ratings:

`=IF(B5="","",B5-AVERAGE($B5:$G5))`

	A	B	C	D	E	F	G
3	Sum of rating	Column Labels ▼					
4	User/ Movie ▼	Just My Luck	Lady in the Water	Snakes on a Plane	Superman Returns	The Night Listener	You Me and Dupree
5	Claudia Puig	3		3.5	4	4.5	2.5
6	Gene Seymour	1.5	3	3.5	5	3	3.5
7	Jack Matthews		3	4	5	3	3.5
8	Lisa Rose	3	2.5	3.5	3.5	3	2.5
9	Mick LaSalle	2	3	4	3	3	2
10	Toby			4.5	4		1
11	Grand Total	9.5	11.5	23	24.5	16.5	15
12							
13	Claudia Puig	-0.5		0	0.5	1	-1
14	Gene Seymour	-1.75	-0.25	0.25	1.75	-0.25	0.25
15	Jack Matthews		-0.7	0.3	1.3	-0.7	-0.2
16	Lisa Rose	0	-0.5	0.5	0.5	0	-0.5
17	Mick LaSalle	-0.833333333	0.166666667	1.166666667	0.166666667	0.166666667	-0.833333333
18	Toby			1.333333333	0.833333333		-2.166666667

2. Calculate the numerator part of the cosine similarity calculation:

`=IF(OR(B$13="",B14=""),"",B$13*B14)`

	A	B	C	D	E	F	G
13	Claudia Puig	-0.5		0	0.5	1	-1
14	Gene Seymour	-1.75	-0.25	0.25	1.75	-0.25	0.25
15	Jack Matthews		-0.7	0.3	1.3	-0.7	-0.2
16	Lisa Rose	0	-0.5	0.5	0.5	0	-0.5
17	Mick LaSalle	-0.833333333	0.166666667	1.166666667	0.166666667	0.166666667	-0.833333333
18	Toby			1.333333333	0.833333333		-2.166666667
19							
20	Gene Seymour	0.875		0	0.875	-0.25	-0.25
21	Jack Matthews			0	0.65	-0.7	0.2
22	Lisa Rose	0		0	0.25	0	0.5
23	Mick LaSalle	0.416666667		0	0.083333333	0.166666667	0.833333333
24	Toby			0	0.416666667		2.166666667

The numerator would be as follows:

`=SUM(B20:G20)`

	A	B	C	D	E	F	G	H	I	J
19										Numerator
20	Gene Seymour	0.875			0	0.875	-0.25	-0.25		1.25
21	Jack Matthews				0	0.65	-0.7	0.2		0.15
22	Lisa Rose	0			0	0.25	0	0.5		0.75
23	Mick LaSalle	0.416666667			0	0.083333333	0.166666667	0.833333333		1.50
24	Toby				0	0.416666667		2.166666667		2.58

3. Prepare the denominator calculator of cosine similarity:

=IF(B13="","",B13^2)

	A	B	C	D	E	F	G
26							
27	Claudia Puig	0.25		0	0.25	1	1
28	Gene Seymour	3.0625	0.0625	0.0625	3.0625	0.0625	0.0625
29	Jack Matthews		0.49	0.09	1.69	0.49	0.04
30	Lisa Rose	0	0.25	0.25	0.25	0	0.25
31	Mick LaSalle	0.694444444	0.027777778	1.361111111	0.027777778	0.027777778	0.694444444
32	Toby			1.777777778	0.694444444		4.694444444

4. Calculate the final cosine similarity, as follows:

=J20/(SQRT(SUM($B$27:$G$27))*SQRT(SUM(B28:G28)))

	I	J	K	L
19		Numerator		
20		1.25		
21		0.15		
22		0.75		
23		1.50		
24		2.58		
25				
26				
27		Cosine similarity		
28	Gene Seymour	0.31		
29	Jack Matthews	0.06		
30	Lisa Rose	0.47		
31	Mick LaSalle	0.56		
32	Toby	0.61		

We now have a similarity value that is associated between –1 and +1 that gives a score of similarity for a given user.

We have now overcome the issue we faced when we had to consider the ratings given by multiple users in predicting the rating that a given user is likely to give to a movie. Users who are more similar to a given user can now be calculated.

Now the problem of predicting the rating that Claudia is likely to give to the movie *Lady in the Water* movie can be solved in the following steps:

1. Normalize users.

2. Calculate cosine similarity.

3. Calculate the weighted average normalized rating.

Let's say we are trying to predict the rating by using the two most similar users instead of one. We would follow these steps:

1. Identify the two most similar users who have also given a rating to the movie *Lady in the Water*.

2. Calculate the weighted average normalized rating that they gave to the movie.

In this case, Lisa and Mick are the two most similar users to Claudia who have rate *Lady in the Water*. (Note that even though Toby is the most similar user, he has not rated *Lady in the Water* and so we cannot consider him for rating prediction.)

Weighted Average Rating Calculation

Let's look at the normalized rating given and the similarity of the two most similar users:

	Similarity	Normalized rating
Lisa Rose	0.47	−0.5
Mick LaSalle	0.56	0.17

The weighted average rating would now be as follows:

$$(0.47 \times -0.5 + 0.56 \times 0.17) / (0.47 + 0.56) = -0.14$$

Potentially, Claudia's average rating would now be reduced by 0.14 to come up with the predicted rating of Claudia for the movie *Lady in the Water*.

Another way to come up with weighted average rating is based on the percent over average rating, as follows:

	Similarity	Normalized rating	Average rating	% avg rating
Lisa Rose	0.47	−0.5	3	−0.5 / 3 = −0.16
Mick LaSalle	0.56	0.17	2.83	0.17 / 2.83 = 0.06

Weighted average normalized rating percentage would now be as follows:

$$(0.47 \times -0.16 + 0.56 \times 0.06) / (0.47 + 0.56) = -0.04$$

Thus, the average rating of Claudia can potentially be reduced by 4% to come up with the predicted rating for the movie *Lady in the Water*.

Choosing the Right Approach

In recommender systems, there is no fixed technique that is proven to always work. This calls for a typical *train, validate, and test* scenario to come up with the optimal parameter combination.

The parameter combination that can be tested is as follows:

- Optimal number of similar users to be considered

- Optimal number of common movies rated together by users before a user is eligible to be considered for similar user calculation

- Weighted average rating calculation approach (based on percentage or on absolute value)

We can iterate through multiple scenarios of various combinations of the parameters, calculate the test dataset accuracy, and decide that the combination that gives the least error rate is the optimal combination for the given dataset.

Calculating the Error

There are multiple ways of calculating, and the preferred method varies by business application. Let's look at two cases:

- Mean squared error (MSE) of all predictions made on the test dataset

- Number of recommended items that a user bought in the next purchase

Note that although MSE helps in building the algorithm, in practice we might be measuring our model's performance as a business-related outcome, as in the second case.

Issues with UBCF

One of the issues with user-based collaborative filtering is that every user has to be compared with every other user to identify the most similar user. Assuming there are 100 customers, this translates into the first user being compared to 99 users, the second user being compared to 98 users, the third to 97, and so on. The total comparisons here would be as follows:

$$99 + 98 + 97 + \ldots + 1 + 0 = 99 \times (99 + 1) / 2 = 4950$$

For a *million* customers, the total number of comparisons would look like this:

$$999{,}999 \times 1{,}000{,}000 / 2 = \sim\!500{,}000{,}000{,}000$$

That's around *500 billion comparisons.* The calculations show that the number of comparisons to identify the most similar customer increases exponentially as the number of customers increases. In production, this becomes a problem because if every user's similarity with every other user needs to be calculated every day (since user preferences and ratings get updated every day based on the latest user data), one would need to perform ~500 billion comparisons every day.

To address this problem, we can consider item-based collaborative filtering instead of user-based collaborative filtering.

Item-Based Collaborative Filtering

Given that the number of computations is an issue in UBCF, we will modify the problem so that we observe the similarity between *items* and not users. The idea behind *item-based collaborative filtering* (*IBCF*) is that two items are similar if the ratings that they get from the same users are similar. Given that IBCF is based on items and not on user similarity, it doesn't have the problem of performing billions of computations.

Let's assume that there are a total of 10,000 movies in a database and 1 million customers attracted to the site. In this case, if we perform UBCF, we would be performing ~500 billion similarity calculations. But using IBCF, we would be performing 9,999 × 5,000 = ~ 50 million similarity calculations.

We can see that the number of similarity calculations increases exponentially as the number of customers grows. However, given that the number of items (movie titles, in our case) is not expected to experience the same growth rate as the number of customers, in general IBCF is less computationally sensitive than UBCF.

The way in which IBCF is calculated and the techniques involved are very similar to UBCF. The only difference is that we would work on a transposed form of the original movie matrix we saw in the previous section. This way, the rows are not of users, but of movies.

Note that although IBCF is better than UBCF in terms of computation, the number of computations is still very high.

Implementing Collaborative Filtering in R

In this section, we will look at the functions used to implement UBCF in R. I implemented the functions available in the `recommenderlab` package in the following code, but in practice it is recommended that you build a recommendation function from scratch to customize for the problem in hand (the code is available as "UBCF.R" in github):

```
# Import data and required packages
t=read.csv("D:/book/Recommender systems/movie_rating.csv")
library(reshape2)
library(recommenderlab)
# Reshape data into a pivot format
t2=acast(t,critic~title)
t2
# Convert it to a matrix
R<-as.matrix(t2)

# Convert R into realRatingMatrix structure
# realRatingMatrix is a recommenderlab sparse-matrix like data structure

r<-as(R,"realRatingMatrix")

# Implement the UBCF method
rec=Recommender(r[1:nrow(r)],method="UBCF")

# Predict the missing rating
recom<-predict(rec,r[1:nrow(r)],type="ratings")
str(recom)
```

In this code, we have reshaped our data so that it can be converted into a `realRatingMatrix` class that gets consumed by the Recommender function to provide the missing value predictions.

Implementing Collaborative Filtering in Python

We used a package in coming up with predictions in R, but for Python we will hand-build a way to come up with predicting the rating a user is likely to give. In the following code, we will create a way to predict the rating that Claudia is likely to give for the *Lady in the Water* movie by considering only the most similar user to Claudia (the code is available as "UBCF.ipynb" in github).

1. Import the dataset:

```python
import pandas as pd
import numpy as np
t=pd.read_csv("D:/book/Recommender systems/movie_rating.csv")
```

2. Convert the dataset into a pivot table:

```python
t2 = pd.pivot_table(t,values='rating',index='critic',columns='title')
```

3. Reset the index:

```python
t3 = t2.reset_index()
t3=t3.drop(['critic'],axis=1)
```

4. Normalize the dataset:

```python
t4=t3.subtract(np.mean(t3,axis=1),axis=0)
```

5. Drop the rows that have a missing value for *Lady in the Water*:

```python
t5=t4.loc[t4['Lady in the Water'].dropna(axis=0).index]
t6=t5.reset_index()
t7=t6.drop(['index'],axis=1)
```

6. Calculate the distance of every other user to Claudia:

```python
x=[]
for i in range(t7.shape[0]):
    x.append(np.mean(np.square(t4.loc[0]-t7.loc[i])))
t6.loc[np.argmin(x)]['Lady in the Water']
```

7. Calculate the predicted rating of Claudia:

```python
np.mean(t3.loc[0]) * (1+(t6.loc[np.argmin(x)]['Lady in
the Water']/np.mean(t3.loc[3])))
```

Working Details of Matrix Factorization

Although user-based or item-based collaborative filtering methods are simple and intuitive, matrix factorization techniques are usually more effective because they allow us to discover latent features underlying the interactions between users and items.

In matrix factorization, if there are U users, each user is represented in K columns, thus we have a $U \times K$ user matrix. Similarly, if there are D items, each item is also represented in K columns, giving us a $D \times K$ matrix.

A matrix multiplication of the user matrix and the transpose of the item matrix would result in the $U \times D$ matrix, where U users may have rated on some of the D items.

The K columns could essentially translate into K features, where a higher or lower magnitude in one or the other feature could give us an indication of the type or genre of the item. This gives us the ability to know the features that a user would give a higher weightage to or the features that a user might not like. Essentially, matrix factorization is a way to represent users and items in such a way that the probability of a user liking or purchasing an item is high if the features that correspond to an item are the features that the user gives a higher weightage to.

We'll see how matrix factorization works through an example. Let's assume that we have a matrix of users (U) and movies (D), as follows (the dataset is available as "matrix factorization example.xlsx" in github):

User	Movies	Actual
1	1	5
1	2	3
1	3	
1	4	1
2	1	4
2	2	
2	3	
2	4	1
3	1	1
3	2	1

(*continued*)

315

User	Movies	Actual
3	3	
3	4	5
4	1	1
4	2	
4	3	
4	4	4
5	1	
5	2	1
5	3	5
5	4	4

Our task is to predict the missing values in the Actual column, which indicate that the user has not rated the movie yet.

In this scenario, the math of matrix factorization works out as follows:

1. Initialize the values of P matrix randomly, where P is a $U \times K$ matrix. We'll assume a value of $k = 2$ for this example.

    A better way of randomly initializing the values is by limiting the values to be between 0 and 1.

In this scenario, the matrix of P will be a 5 × 2 matrix, because $k = 2$ and there are 5 users:

	P matrix	
	Factor 1	Factor 2
User1	0.44	0.52
User2	0.57	0.11
User3	0.53	0.27
User4	0.82	0.04
User5	0.39	0.74

2. Initialize the values of Q matrix randomly, again where Q is a $K \times D$ matrix—that is, a 2 × 4 matrix, because there are four movies, as shown in the first table.

The Q matrix would be as follows:

	Q matrix			
	Movie1	Movie2	Movie3	Movie4
Factor 1	0.81	0.03	0.79	0.71
Factor 2	0.34	0.84	0.09	0.49

3. Calculate the value of the matrix multiplication of P × Q matrix.

    Note that the Prediction column in the following is calculated by
    the matrix multiplication of P matrix and Q matrix (I will discuss
    the Constraint column in the next step):

=C4*F4+D4*F5

	A	B	C	D	E	F	G	H	I
1									
2			P matrix				Q matrix		
3			Factor 1	Factor 2		Movie1	Movie2	Movie3	Movie4
4		User1	0.07	0.68	Factor 1	0.42	0.79	0.18	0.13
5		User2	0.4	0.34	Factor 2	0.11	0.94	0.52	0.79
6		User3	0.27	0.23					
7		User4	0.77	0.8					
8		User5	0.97	0.36					
9									
10			User	Movies	Constraint	Prediction	Actual		
11			1	1	0	0.1042	5		
12			1	2	0	0.6945	3		
13			1	3	0	0.3662			
14			1	4	0	0.5463	1		
15			2	1	0	0.2054	4		
16			2	2	0	0.6356			
17			2	3	0	0.2488			
18			2	4	0	0.3206	1		
19			3	1	0	0.1387	1		
20			3	2	0	0.4295	1		
21			3	3	0	0.1682			
22			3	4	0	0.2168	5		
23			4	1	0	0.4114	1		
24			4	2	0	1.3603			
25			4	3	0	0.5546			
26			4	4	0	0.7321	4		
27			5	1	0	0.447			
28			5	2	0	1.1047	1		
29			5	3	0	0.3618	5		
30			5	4	0	0.4105	4		

4. Specify the optimization constraints.

    The predicted value (the multiplication of each element of the
    two matrices) should ideally be equal to the ratings of the big
    matrix. The error calculation is based on the typical squared error

calculation and is done as follows (note that the weight values in P and matrices have varied because they are random numbers and are initialized using the randbetween function, which changes values every time Enter is pressed in Excel):

`=(F11-G11)^2`

	A	B	C	D	E	F	G	H	I
1									
2			P matrix				Q matrix		
3			Factor 1	Factor 2		Movie1	Movie2	Movie3	Movie4
4		User1	0.53	0.72	Factor 1	0.51	0.07	0.99	0.6
5		User2	0.87	0	Factor 2	0.45	0.8	0.81	0.81
6		User3	0.12	0.85					
7		User4	0.75	0.61					
8		User5	0.26	0.81					
9									
10			User	Movies	Constraint	Prediction	Actual	error	
11			1	1	0	0.5943	5	19.410192	
12			1	2	0	0.6131	3	5.6972916	
13			1	3	0	1.1079		1.2274424	
14			1	4	0	0.9012	1	0.0097614	
15			2	1	0	0.4437	4	12.64727	
16			2	2	0	0.0609		0.0037088	
17			2	3	0	0.8613		0.7418377	
18			2	4	0	0.522	1	0.228484	
19			3	1	0	0.4437	1	0.3094697	
20			3	2	0	0.6884	1	0.0970946	
21			3	3	0	0.8073		0.6517333	
22			3	4	0	0.7605	5	17.97336	
23			4	1	0	0.657	1	0.117649	
24			4	2	0	0.5405		0.2921403	
25			4	3	0	1.2366		1.5291796	
26			4	4	0	0.9441	4	9.3385248	
27			5	1	0	0.4971		0.2471084	
28			5	2	0	0.6662	1	0.1114224	
29			5	3	0	0.9135	5	16.699482	
30			5	4	0	0.8121	4	10.162706	
31						Overall error		97.495859	

- *Objective*: Change the randomly initialized values of P and Q matrices to minimize overall error.

- *Constraint*: No prediction can be greater than 5 or less than 1.

The preceding objective and constraint can be specified as an optimization scenario in Solver, as follows:

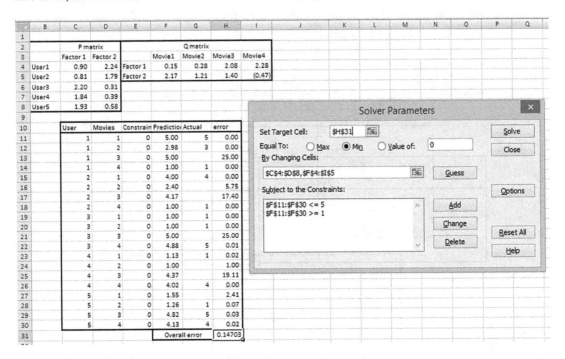

Note that once we optimize for the given objective and constraint, the optimal values of weights in P and Q matrices are arrived at and are as follows:

		P matrix			Q matrix			
		Factor 1	Factor 2		Movie1	Movie2	Movie3	Movie4
4	User1	0.90	2.24	Factor 1	0.15	0.29	2.08	2.28
5	User2	0.81	1.79	Factor 2	2.17	1.21	1.40	(0.47)
6	User3	2.20	0.31					
7	User4	1.84	0.39					
8	User5	1.93	0.58					

INSIGHTS ON P AND Q MATRICES

In P matrix, user 1 and user 2 have similar weightages for factors 1 and 2, so they can potentially be considered to be similar users.

Also, the way in which user 1 and 2 have rated movies is very similar—the movies that user 1 rated highly have a high rating from user 2 as well. Similarly, the movies that user 1 rated poorly also had low ratings from user 2.

The same goes for the interpretation for Q matrix (the movie matrix). Movie 1 and movie 4 have quite some distance between them. We can also see that, for a majority of users, if the rating given for movie 1 is high, then movie 4 got a low rating, and vice versa.

Implementing Matrix Factorization in Python

Notice that the P matrix and Q matrix are obtained through Excel's Solver, which essentially runs gradient descent in the back end. In other words, we are deriving the weights in a manner similar to the neural networks–based approach, where we are trying to minimize the overall squared error.

Let's look at implementing matrix factorization in keras for the following dataset (the code is available as "matrix factorization.ipynb" in github):

User	Movies	Actual
1	4	1
2	4	1
3	1	1
3	2	1
4	1	1
5	2	1
1	2	3
2	1	4
4	4	4

(continued)

User	Movies	Actual
5	4	4
1	1	5
3	4	5
5	3	5

```
# Import the required packages and dataset
import pandas as pd
ratings= pd.read_csv('/content/datalab/matrix_factorization_keras.csv')

# Extract the unique users
users = ratings.User.unique()

# Extract the unique movies
articles = ratings.Movies.unique()
# Index each user and article
userid2idx = {o:i for i,o in enumerate(users)}
articlesid2idx = {o:i for i,o in enumerate(articles)}

# Apply the index created to the original dataset
ratings.Movies = ratings.Movies.apply(lambda x: articlesid2idx[x])
ratings.User = ratings.User.apply(lambda x: userid2idx[x])

# Extract the number of unique users and articles
n_users = ratings.User.nunique()
n_articles = ratings.Movies.nunique()

# Define the error metric
import keras.backend as K
def rmse(y_true,y_pred):
    score = K.sqrt(K.mean(K.pow(y_true - y_pred, 2)))
    return score

# Import relevant packages
from keras.layers import Input, Embedding, Dense, Dropout, merge, Flatten
from keras.models import Model
```

The function Embedding helps in creating vectors similar to the way we converted a word into a lower-dimensional vector in Chapter 8.

Through the following code, we would be able to create the initialization of P matrix and Q matrix:

```
def embedding_input(name,n_in,n_out):
    inp = Input(shape=(1,),dtype='int64',name=name)
    return inp, Embedding(n_in,n_out,input_length=1)(inp)
n_factors = 2
user_in, u = embedding_input('user_in', n_users, n_factors)
article_in, a = embedding_input('article_in', n_articles, n_factors)

# Initialize the dot product between user matrix and movie matrix
x = merge.dot([u,a],axes=2)
x=Flatten()(x)

# Initialize the model specification
from keras import optimizers
model = Model([user_in,article_in],x)
sgd = optimizers.SGD(lr=0.01)
model.compile(sgd,loss='mse',metrics=[rmse])
model.summary()

# Fit the model by specifying inputs and output
model.fit([ratings.User,ratings.Movies], ratings.Actual, nb_epoch=1000,
batch_size=13)
```

Now that the model is built, let's extract the weights of the User and Movie matrices (P and Q matrices):

```
# User matrix
model.get_weights()[0]

array([[ 1.3606772,  1.6446526],
       [ 1.0029647,  1.3566955],
       [-1.4094337,  1.6923367],
       [-1.0425398,  1.4308459],
       [-1.2908196,  1.4010738]], dtype=float32)
```

```
# Movie matrix
model.get_weights()[1]

array([[-1.3724959,  1.7509296],
       [ 1.5061257,  1.8171026],
       [ 0.6696424,  1.235024 ],
       [-1.7306372,  1.8775703]], dtype=float32)
```

Implementing Matrix Factorization in R

Although matrix factorization can be implemented using the kerasR package, we will use the recommenderlab package (the same one we worked with for collaborative filtering).

The following code implements matrix factorization in R (available as "matrix factorization.R" in github):

1. Import the relevant packages and dataset:

```
# Matrix factorization
t=read.csv("D:/book/Recommender systems/movie_rating.csv")
library(reshape2)
library(recommenderlab)
```

2. Pre-process the data:

```
t2=acast(t,critic~title)
t2
# Convert it as a matrix
R<-as.matrix(t2)
# Convert R into realRatingMatrix data structure
# RealRatingMatrix is a recommenderlab sparse-matrix like data-
structure
r <- as(R, "realRatingMatrix")
```

3. Use the funkSVD function to build the matrix factors:

```
fsvd <- funkSVD(r, k=2,verbose = TRUE)
p <- predict(fsvd, r, verbose = TRUE)
p
```

Note that the object p constitutes the predicted ratings of all the movies across all the users.

The object fsvd constitutes the user and item matrices, and they can be obtained with the following code:

```
str(fsvd)
```

```
> str(fsvd)
List of 3
 $ U         : num [1:6, 1:2] 1.61 1.72 1.74 1.53 1.46 ...
 $ V         : num [1:6, 1:2] 0.971 1.178 1.918 2.081 1.53 ...
 $ parameters:List of 6
  ..$ k              : num 2
  ..$ gamma          : num 0.015
  ..$ lambda         : num 0.001
  ..$ min_epochs     : num 50
  ..$ max_epochs     : num 200
  ..$ min_improvement: num 1e-06
 - attr(*, "class")= chr "funkSVD"
```

The user matrix can thus be accessed as fsvd$U, and the item matrix by fsvd$V. The parameters are the learning rate and the epochs parameters we learned about in Chapter 7.

Summary

In this chapter, you have learned the following:

- The major techniques used to provide recommendations are collaborative filtering and matrix factorization.

- Collaborative filtering is extremely prohibitive in terms of the large number of computations.

- Matrix factorization is less computationally intensive and in general provides better results.

- Ways to build matrix factorization and collaborative filtering algorithms in Excel, Python, and R

CHAPTER 14

Implementing Algorithms in the Cloud

Sometimes the amount of computation required to carry out a task can be enormous. This typically happens when there is a large dataset that has a size greater than the typical RAM size of a machine. It can also typically happen when the required processing on the data is huge.

In such cases, it is a good idea to switch to *cloud-based analysis*, which can help scale up a larger RAM size quickly. It can also avoid the need to purchase extended RAM to resolve an issue that might not occur very frequently. However, there is a cost to use cloud services, and certain configurations are more costly than others. You need to be careful when choosing a configuration and be disciplined so that you know when to stop using the cloud service.

The three major cloud providers are as follows:

- Google Cloud Platform (GCP)

- Microsoft Azure

- Amazon Web Services (AWS)

In this chapter, we will work towards setting up a virtual machine in all the three cloud platforms. Once the instance is setup, we will learn about accessing Python and R on cloud.

Google Cloud Platform

GCP can be accessed at `https://cloud.google.com`. Once you set up an account, you use the console to create a project. In the console, click Compute Engine and then click VM instances, as shown in Figure 14-1 (*VM* stands for *virtual machine*).

© V Kishore Ayyadevara 2018
V. K. Ayyadevara, *Pro Machine Learning Algorithms*, https://doi.org/10.1007/978-1-4842-3564-5_14

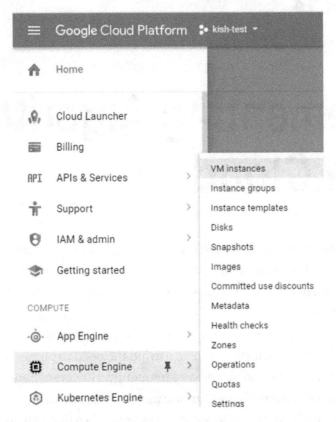

Figure 14-1. *Selecting the VM option*

Click Create to create a new VM instance. You will see a screen like Figure 14-2.

← Create an instance

Name ⓘ

instance-4

Zone ⓘ

us-central1-c

$24.67 per month estimated
Effective hourly rate $0.034 (730 hours per month)

Machine type
Customise to select cores, memory and GPUs.

⌄ Details

| 1 vCPU ▾ | 3.75 GB memory | Customise |

Container ⓘ
☐ Deploy a container image to this VM instance. Learn more

Figure 14-2. *Options to create an instance*

Depending on the dataset size, customize the machine type with the required cores, memory, and whether you need a GPU.

Next, select the operating system of your choice (Figure 14-3).

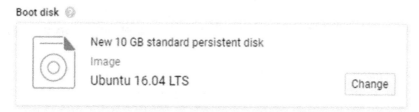

Boot disk ⓘ

New 10 GB standard persistent disk
Image
Ubuntu 16.04 LTS

Change

Figure 14-3. *Selecting the OS*

We will perform a few operations on PuTTY. You can download PuTTYgen from www.ssh.com/ssh/putty/windows/puttygen. Open the program and click Generate to generate a public/private key pair. A key will be generated, as shown in Figure 14-4.

```
PuTTY Key Generator                                    ?   ×

File   Key   Conversions   Help

 Key
  Public key for pasting into OpenSSH authorized_keys file:
  ssh-rsa AAAAB3NzaC1yc2EAAAABJQAAAQEAuHar/vTtYgElgnfJXfVYZf
  +0xJtS/3OtuGDWj1frcuOnEssxKlh6J5Q5dHnPBZ/fi0hB5ypCXOfbQVu/ENYHuBdArSL
  5o8t8oUkMq49BzDug4nlaZ
  +yO36gD8erJSFyqfn5JWEIXRpWkT4MAryqqOOd0f1OAHZOM1OEAb/l8wm2tJYaPy4
  78DWkfyzveUjvUfbkSrhLL5itv19CE9zhwATNr3mRNjIJMFiMehyjMEz9Fonle0lh3tprFqc

  Key fingerprint:      ssh-rsa 2048 0d:7a:21:92:47:54:ce:33:95:17:33:63:0b:4e:f3:b5
  Key comment:          rsa-key-20180118
  Key passphrase:
  Confirm passphrase:

 Actions
  Generate a public/private key pair                              Generate
  Load an existing private key file                                 Load
  Save the generated key           Save public key        Save private key

 Parameters
  Type of key to generate:
  ● RSA       ○ DSA        ○ ECDSA       ○ ED25519      ○ SSH-1 (RSA)
  Number of bits in a generated key:                     2048
```

Figure 14-4. *Generating a public/private key pair in PuTTYgen*

Click "Save private key" to save the private key. Copy the public key at the top and paste it into the SSH Keys box on GCP, as shown in Figure 14-5.

Figure 14-5. *Pasting in the key*

Click Create. That should create a new instance for you. It should also give you the IP address corresponding to the instance. Copy the IP address and paste it in PuTTY under Host Name.

Click SSH in the left pane, click Auth, and browse to the location where you saved the private key, as shown in Figure 14-6.

Figure 14-6. *The Auth options*

Enter the login name as the "Key comment" entry shown in PuTTYgen when you were generating public and private keys back in Figure 14-4. You should now be logged into the Google Cloud machine.

Type **python**, and you should be able to run Python scripts.

Be sure to delete the instance as soon as you are done with your work. Otherwise the service may still be billing you.

Microsoft Azure Cloud Platform

Creating a virtual machine instance in Microsoft Azure is very similar to the way it is done in GCP. Visit https://azure.microsoft.com and set up an account.

Create an account in Azure and log in. Then click "Virtual machines," as shown in Figure 14-7.

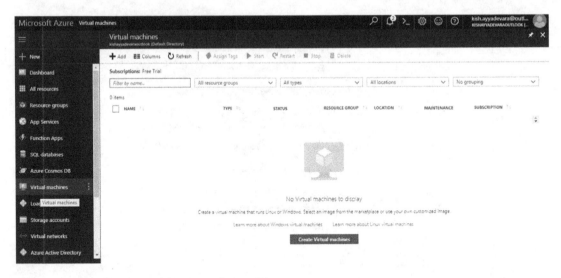

Figure 14-7. *Microsoft's Virtual machines page*

Click Add and then do the following:

1. Select the machine needed—in our case, I'm selecting Ubuntu Server 16.04 LTS.

2. Click the default Create button.

3. Enter the machine-level details in basic configuration settings.

4. Select the size of virtual machine needed.

5. Configure the optional features.

6. Finally, create the instance.

Once the instance is created, the dashboard provides the IP address corresponding to the instance, as shown in Figure 14-8.

Figure 14-8. *The IP address you need*

Open PuTTY (see the previous section for more on downloading and launching it) and connect to the instance using the IP address, either by entering the password (if you selected the password option while creating the instance) or by using the private key.

You can connect to the instance and open Python using PuTTY in a similar way as we did in the previous section on GCP.

Amazon Web Services

In this section, we will sign up with Amazon Web Services. Got to https://aws.amazon.com and create an account.

Click "Launch a virtual machine," as shown in Figure 14-9.

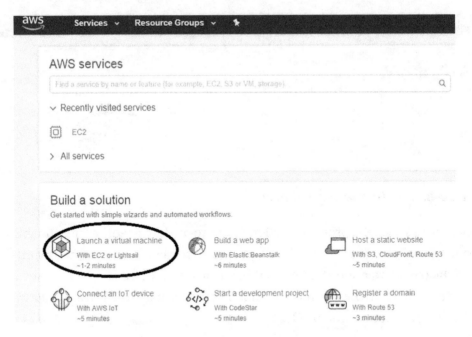

Figure 14-9. *Launching a virtual machine in AWS*

On the next screen, click "Get started." Name your instance and select the required attributes, as shown in Figure 14-10.

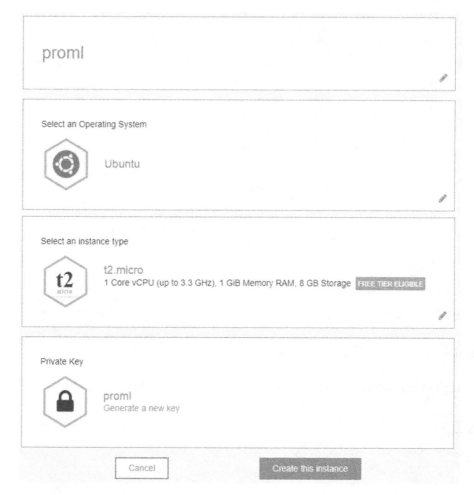

Figure 14-10. Setting up your instance

Download the .pem file and click "Create this instance." Then click "Proceed to console."

In the meantime,

1. Open PuTTYgen.

2. Load the .pem file.

3. Convert it into a .ppk file.

4. Save the private key, as shown in Figure 14-11.

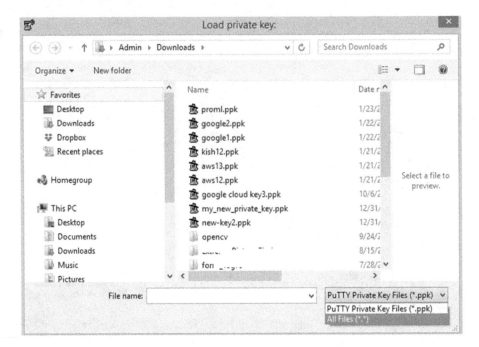

Figure 14-11. *Saving the private key*

Go back to the AWS console, where the screen looks like Figure 14-12.

Figure 14-12. *The AWS console*

Click the Connect button. Note the example given in the pop-up that appears (Figure 14-13).

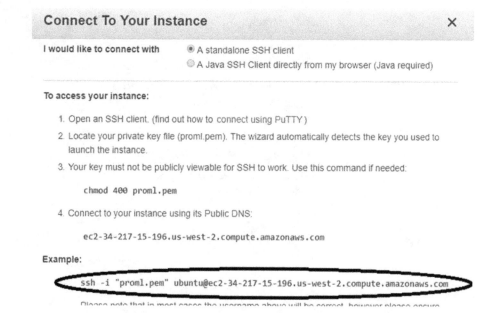

Figure 14-13. *The example*

In the highlighted part of Figure 14-13, the string after the @ is the host name. Copy it, open PuTTY, and paste the host name into the Host Name box in the PuTTY Configuration screen, as shown in Figure 14-14.

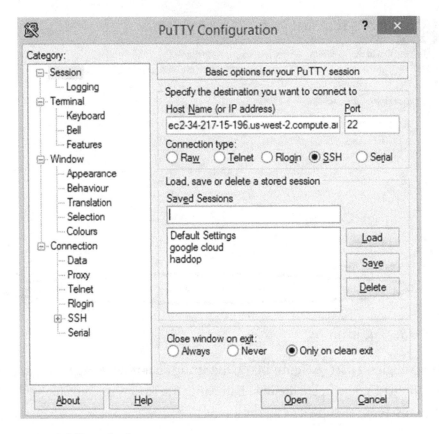

Figure 14-14. *Adding the host name*

Back in Figure 14-13, the word just before the @ is the username. Type it into the "Auto-login username" box in PuTTY after you select "Data" in panel in the left, as shown in Figure 14-15.

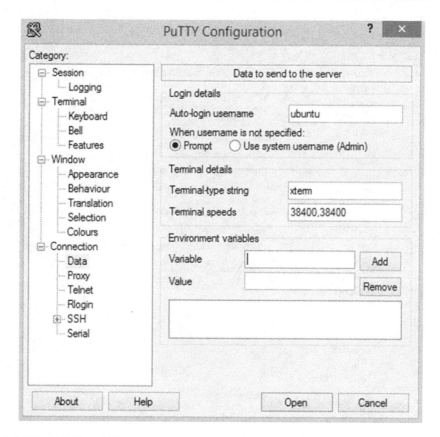

Figure 14-15. *Adding the username*

Click SSH to expand it, click Auth, and browse to the .ppk file created earlier. Click Open, as shown in Figure 14-16.

Figure 14-16. *Setting the private key*

Now you should be able to run Python on AWS.

Transferring Files to the Cloud Instance

You can transfer files from your local machine to your cloud instance in all three platforms using WinSCP. If you don't already have it installed, download WinSCP from www.winscp.net and install it. Open WinSCP and you should see a login screen similar to Figure 14-17.

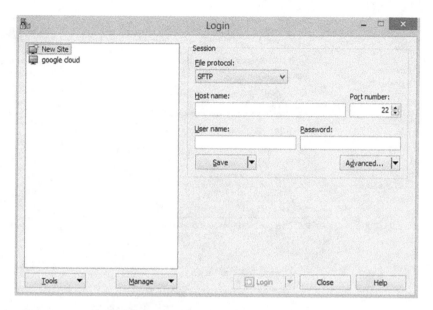

Figure 14-17. *The WinSCP login screen*

Enter the host name and username similar to how you entered them in PuTTY. In order to enter the .ppk file details, click the Advanced button.

Click Authentication in the SSH section and provide the location of the .ppk file, as shown in Figure 14-18.

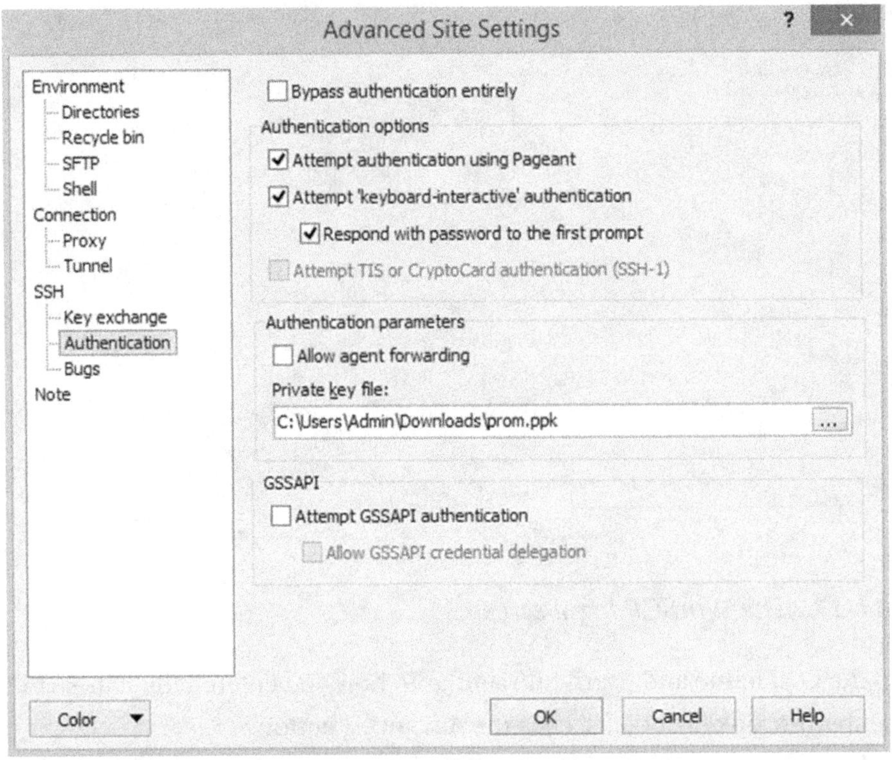

Figure 14-18. *Setting the private key*

Click OK and then click Login.

Now you should be able to transfer files from your local machine to the virtual instance.

Another way to transfer files is to upload them into some other cloud storage (for example, Dropbox), obtain a link for the location of the file, and download it to the virtual instance.

Running Instance Jupyter Notebooks from Your Local Machine

You can run Jupyter Notebooks from your local machine by running the following code on Linux instances in any of GCP, AWS, or Azure:

```
sudo su
wget http://repo.continuum.io/archive/Anaconda3-4.1.1-Linux-x86_64.sh
bash Anaconda3-4.1.1-Linux-x86_64.sh
```

```
jupyter notebook --generate-config
vi jupyter_notebook_config.py
```

Insert the following code by pressing the I key:

```
c = get_config()
c.NotebookApp.ip = '*';
c.NotebookApp.open_browser = False
c.NotebookApp.port = 5000
```

Press Escape, type **:wq**, and press Enter.

Type the following:

```
sudo su
jupyter-notebook --no-browser --port=5000
```

Once the Jupyter Notebook opens, go to a browser on the local machine and type the IP address of the virtual instance, along with the port number into the address bar (make sure that the firewall rules are configured to open port 5000). For example, if the IP address is http://35.188.168.71, then type **http://35.188.168.71:5000** into the browser's address bar at the top of the screen.

You should be able to see the Jupyter environment on your local machine that is connected to the virtual instance (Figure 14-19).

Figure 14-19. *The Jupyter environment*

Installing R on the Instance

R does not come installed by default on the instance. You can install R in Linux as follows:

```
sudo apt-get update
sudo apt-get install r-base
```

Now type **R** in your terminal:

```
R
```

You should now be able to run R code in the virtual instance.

Summary

In this chapter, you learned the following:

- How to set up and open a virtual instance on the three major cloud platforms
- How to run Python/R on the three platforms
- How to transfer a file into the cloud environment

APPENDIX

Basics of Excel, R, and Python

This chapter introduces the three tools mentioned often in this book: Microsoft Excel and the two programming languages R and Python.

Basics of Excel

In Microsoft Excel, each cell is represented by numbers in rows and letters in columns.

For example, the following highlighted cell is the cell D4:

Cells can be referenced by specifying the = symbol followed by the cell we are trying to refer to. For example, if we want the cell D4 to reflect the value in cell A1, we would type it in as follows:

© V Kishore Ayyadevara 2018
V. K. Ayyadevara, *Pro Machine Learning Algorithms*, https://doi.org/10.1007/978-1-4842-3564-5

Pressing F2 gets you to the formula corresponding to a cell.

We can do multiple manipulations on top of a given cell value with the various functions built-in to Excel. For example, here's how to equate the cell value of D4 to the exponential of the cell value of A1:

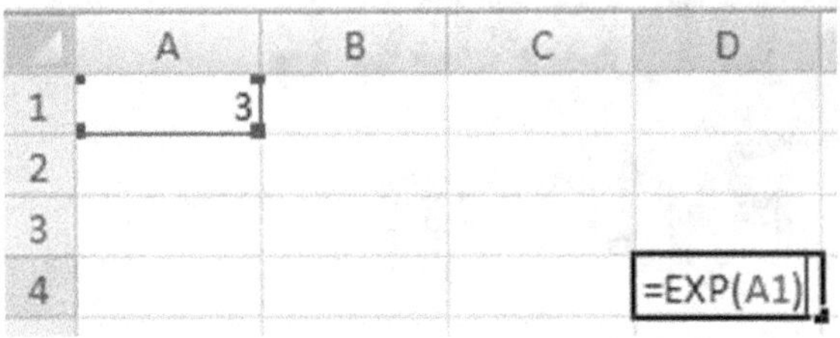

◢	A	B	C	D
1	3			
2				
3				
4				=EXP(A1)

Excel provides an optimization tool that comes in handy in various techniques discussed in this book, called the Solver. Excel Solver is an add-in that must be installed. Once installed, it's available in the Data tab in the Excel ribbon at the top.

A typical Solver looks like Figure A-1.

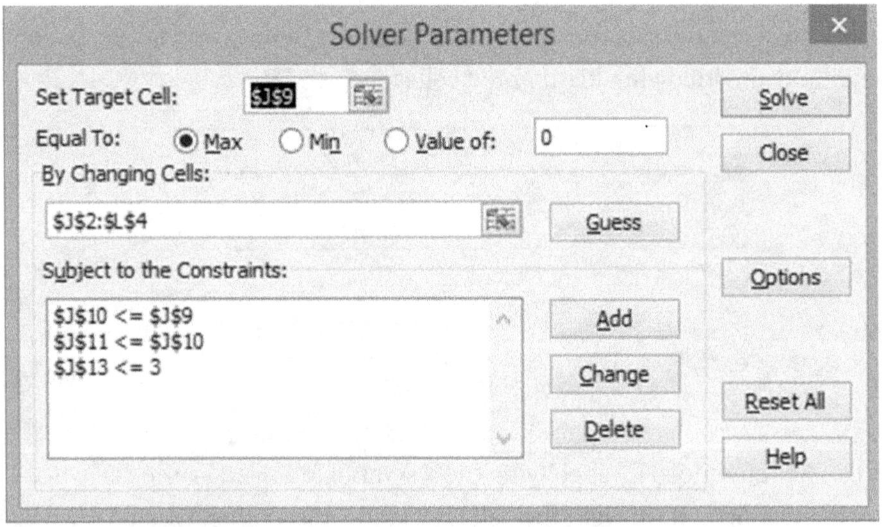

Figure A-1. *A typical Solver*

In the Set Target Cell part of Solver, you can specify the target that needs to be worked on.

In the Equal To section, you specify the objective—whether you want to minimize the target, maximize it, or set it to a value. This comes in very handy in scenarios where the target is the error value and we want to minimize the error.

The next section, "By Changing Cells", specifies the cells that can be changed to achieve the objective.

Finally, the Subject to the Constraints section specifies all the constraints that govern the objective.

Clicking the Solve button gives us the optimal cell values that achieve our objective.

Solver works using multiple algorithms, all of which are based on *back-propagation* (a technique discussed in detail in Chapter 7).

There are a lot more functionalities in Excel that can be very helpful, but for the purposes of this book—to show you how the algorithms work—you are in good shape if you understand Solver and cell linkages well.

Basics of R

The R programming language is an offshoot of a programming language called S. R was developed by Ross Ihaka and Robert Gentleman from the University of Auckland, New Zealand. It was primarily adopted by statisticians and is now the de facto standard language for statistical computing.

- *R is data analysis software*: Data scientists, statisticians, analysts, and others who need to make sense of data use R for statistical analysis, predictive modeling, and data visualization.

- *R is a programming language*: You do data analysis in R by writing scripts and functions in the R programming language. R is a complete, interactive, object-oriented language. The language provides objects, operators, and functions that make the process of exploring, modeling, and visualizing data a natural one. Complete data analyses can often be represented in just a few lines of code.

- *R is an environment for statistical analysis*: Available in the R language are functions for virtually every date manipulation, statistical model, or chart that the data analyst could ever need.

- *R is an open source software project*: Not only does this mean that you can download and use R for free, but the source code is also open for inspection and modification to anyone who wants to see how the methods and algorithms work under the hood.

Downloading R

R works on many operating systems, including Windows, Macintosh, and Linux. Because R is free software, it is hosted on many different servers (mirrors) around the world and can be downloaded from any of them. For faster downloads, you should choose a server close to your physical location. A list of all available download mirrors is available at www.r-project.org/index.html. Click Download R on the front page to choose your mirror for downloading.

You may notice that many of the download URLs include the letters CRAN. *CRAN* stands for the Comprehensive R Archive Network, and it ensures that you have the most recent version of R.

Once you have chosen a mirror, at the top of your screen you should see a list of the versions of R for each operating system. Choose the R version that works on your operating system (also, you should download base R), and then click the download link to download.

Installing and Configuring RStudio

RStudio is an integrated development environment (IDE) dedicated to R development.

RStudio requires a pre-installed R instance, and in RStudio config, an R version must be set (usually it is auto-set by RStudio when it is installed). RStudio is a more user-friendly version of R compared to the native R version.

1. Go to www.rstudio.com/products/rstudio/download/.

2. Click the Download RStudio Desktop button.

3. Select the installation file for your system.

4. Run the installation file.

5. RStudio will be installed on your system. It normally detects your latest installed R version automatically. Ideally, you should be able to use R from within RStudio without extra configuration.

Getting Started with RStudio

RStudio displays the main windows shown in Figure A-2.

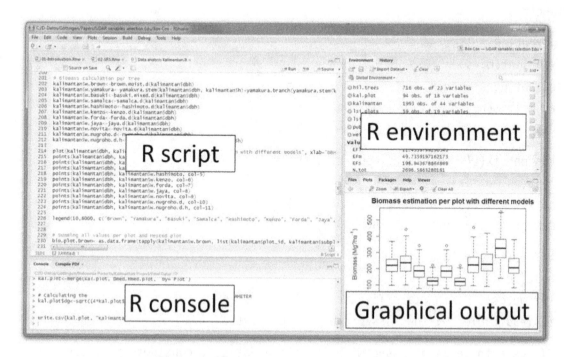

Figure A-2. *RStudio's main windows*

You can perform various functions in R, as follows (code available in github as "R basics.R"):

```
#Basic calculations
1+1
2*2
#Logical operators
1>2
1<2
1&0
1|0
#Creating a vector
n = c(2, 3, 5,6,7)
s = c("aa", "bb", "cc", "dd", "ee")
```

```
s
s = c( "bb", 1)
s
a=c(n,s)

#Creating a list
x = list(n, s)
#Creating a matrix
as.matrix(n)
as.matrix(s)

#as.matrix(c(n,s),nrow=5,ncol=2,byrow=TRUE)
help(as.matrix)
matrix(c(1,2,3, "a","b","c"), nrow = 3, ncol = 2)
matrix(c(n,s), nrow = 5, ncol = 2)

# Importing datasets
help(read)
t=read.csv("D:/in-class/Titanic.csv")
t2=read.table("D:/in-class/credit_default.txt")

#In case there is an issue with importing dataset, consider specifying
quote=, as below
t3=read.csv("D:/in-class/product_search.csv",quote="\"")

# Structure of dataset
t=read.csv("D:/in-class/Titanic.csv")
str(t)  # gives us quite a few information about the dataset
class(t) # typically, the imported datasets as always imported as data.
frame objects
dim(t) # dimension of an abject (data.frame)
nrow(t) # number of rows
ncol(t) # number of columns
colnames(t) # specify column names
class(t$Survived) # class of a variable
head(t) # gives us the first few rows of a dataset
t[1,1] # gives us the value in first row & first column. Note the syntax:
[rows,columns]
```

```
t[1:3,1] # gives us the values in first 3 rows of 1st column
t[c(1,100,500),]
t[1:3,] # gives us the values in first 3 rows of all columns (note that
when we dont specify the filtering condition in column index, the result
includes all columns)
t$Survived[1:3]# gives us the values of first 3 rows of "Survived" variable

t[c(1,4),-c(1,3)] # c() is a function to get a combination of values.
c(1,4) gives us the first & 4th row. -c(1) excludes the first column
t[c(1,4),c("Survived","Fare")] # c() is a function to get a combination of
values. c(1,4) gives us the first & 4th row. -c(1) excludes the first column
t[1:3,"Survived"]
# Data manipulation
t$PassengerId=NULL
summary(t)
t$unknown_age=0
t$unknown_age=ifelse(is.na(t$Age),1,0)  # the syntax here is,
ifelse(condition, value if condition is true, value if condition is false)
# is.na() function helps us in identifying if there any NA values in
dataset. Be sure to remove or impute (replace) NA values within a dataset
unique(t$Embarked) # gives all the unique values within a variable
table(t$Embarked) # table gives a count of all the unique values in dataset
mean(t$Age) # mean as a function
sum(t$Age)
mean(t$Age,na.rm=TRUE) # na.rm=TRUE helps in removing the missing values,
if they exist
t$Age2=ifelse(t$unknown_age==1,mean(t$Age,na.rm=TRUE),t$Age) # one
initializes a new variable within a dataset by using the $ operator & the
new variable name
summary(t) # summary is typically the first step performed after importing
dataset
order(t$Age2) # order function is used to sort dataset. It gives out the
index of rows that a value is sorted to
t=t[order(t$Age2),]  # t is sorted based on the order of age
mean(t$Survived[1:50])
mean(t$Survived[(nrow(t)-50):nrow(t)])
```

```
t2=t[1:50,]
t_male=t[t$Sex=="male",] # one can filter for criterion by specifying the
variable name with a == and the value it is to be filtered to
t_female=t[t$Sex=="female",]

mean(t_male$Survived)
mean(t_female$Survived)

t2=t[t$Sex=="male" & t$Age2<10,]
mean(t2$Survived)
t2=t[t$Sex=="female" | t$Age2<10,]
mean(t2$Survived)

install.packages("sqldf")
library(sqldf)
t3=sqldf("select sex,avg(survived) from t group by sex")
# SQL like filtering or aggregation can be done using the sqldf function
t$age3=ifelse(t$Age2<10,1,0)
t2=t[,c("Age2","Sex","Survived")] # one can filter for the columns of
interest y specifying the c() function with the variables that are needed
help(aggregate)
aggregate(t$Survived,by=list(t$Sex,t$Pclass),sum)
# aggregate function works similar to sqldf where grouping operations can
be done
#seq function is used to generate numbers by a given step size.
seq(0,1,0.2) gives us c(0,0.2,0.4,0.6,0.8,1)
#quantile gives us the values at the various percentiles specified
help(quantile)
summary(t)
t$age3=as.character(t$Age) # as.character function converts a value into a
character variablle
quantile(t$Age2,probs=seq(0,0.5,0.1))[2] # gives us the second value in the
output of quantile function

x=quantile(t$Age2,probs=seq(0,1,0.1))[2]
t2=t[t$Age2<x,]
mean(t2$Survived)
```

```
t$less_than_10=ifelse(t$Age2<x,1,0)
aggregate(t$Survived,by=list(t$Sex,t$less_than_10),mean) # aggregation can
be done over multiple variables by using c() function

t2=t[!t$Age2<x,] # ! is used as a engation statement
mean(t2$Survived)

# Loops
t=read.table("D:/in-class/credit_default.txt")

for(i in 1:3){
  print(i)
}
summary(t)
# a good idea is to note the difference between mean & median values of
variables
mean(t$DebtRatio)
median(t$DebtRatio)
t2=t

# it's a good practcie to test out the code of for loop before looping it
through, by assigning a certain value of i & test out the for loop code
i=2

t2[,i]=ifelse(is.na(t2[,i]),median(t2[,i],na.rm=TRUE),t2[,i])
t2[,i]=ifelse(t2[,i]<median(t2[,i],na.rm=TRUE),"Low","High")
print(aggregate(t2$SeriousDlqin2yrs,by=list(t2[,i]),mean))

for(i in 1:ncol(t2)){
  t2[,i]=ifelse(is.na(t2[,i]),median(t2[,i],na.rm=TRUE),t2[,i])
}

# below is an exercise where we are imputing missing value with median
values & then flagging variables as high when the value is above median
value & low when the value is below median value
for(i in 2:ncol(t2)){
  t2[,i]=ifelse(is.na(t2[,i]),median(t2[,i],na.rm=TRUE),t2[,i])
  t2[,i]=ifelse(t2[,i]<median(t2[,i],na.rm=TRUE),"Low","High")
  print(colnames(t2[i]));
```

```
  print(aggregate(t2$SeriousDlqin2yrs,by=list(t2[,i]),mean))
}
df=data.frame(group=c("a","b"),avg=c(2,2))

#joins
search=read.csv("D:/in-class/product_search.csv",quote="\"")
descriptions=read.csv("D:/in-class/product_descriptions.csv",quote="\"")
summary(search)
colnames(search)
colnames(descriptions)

help(merge)
# in a typical merge function, we have to specify the x (first) table, the
y (second) table to which we are joining the first table
# we would also have to specify the variable based on which we are joining
the datasets using the "by" parameter
# in case the column name of by parameter is different in datasets, we can
use by.x & by.y
# by default merge does an inner join (inner join is when only the values
that are common in both tables are joined)
# all.x=TRUE helps us do a left join (left join is when all the values in x
table are retained even if some of them do not have a match in the second
table)
# all.y = TRUE does a right join where all the values in right (second)
table are retained
# assume that x table has productid as (1,2,3) & right (y) table has
productid as (1,5,6)
# inner join of these two tables gives us the values of only productid =1
(as it is the only one in common)
# left join gives us the information of (1,2,3) however, the info of pid 1
will be full and info of pid 2,3 would be blank, as right table does not
have info about these 2 pids
# right join givs the information of (1,5,6) where we have pid 1 info
completely and info of 5,6 is missing

search_descriptions=merge(search,descriptions,by="product_uid",all.x=TRUE)
search_descriptions1=merge(search,descriptions,by="product_uid",all.y=TRUE)
```

```r
search_descriptions2=merge(descriptions,search,by="product_uid",all.x=TRUE)
nrow(search_descriptions)
nrow(search_descriptions1)
nrow(search_descriptions2)

search_descriptions2$missing_id=ifelse(is.na(search_descriptions2$id),1,0)
sum(search_descriptions2$missing_id)
x=search_descriptions2[search_descriptions2$missing_id==0,]
length(unique(x$Product_uid))

system.time(search_descriptions<-merge(search,descriptions,by="product_
uid",all.x=TRUE))

# note the difference in speed between base "merge" statement & fread/
data.table "merge" statment
install.packages("data.table")
library(data.table)
search=fread("D:/in-class/product_search.csv")
descriptions=fread("D:/in-class/product_descriptions.csv")

system.time(descriptions<-read.csv("D:/in-class/product_descriptions.csv"))
system.time(descriptions<-fread("D:/in-class/product_descriptions.csv"))

write.csv(search,"D:/in-class/search_output.csv",row.names=FALSE)

help(merge)

search_descriptions=merge(search,descriptions,by="product_uid",all.x=TRUE)
system.time(search_descriptions<-merge(search,descriptions,by="product_
uid",all.x=TRUE))

# writing custom functions
square = function(x) {x*x}

square(13.5)
square("two")

addition = function(x,y) {x+y}

tt=as.data.frame(quantile(t$Age2,probs=seq(0,1,0.1)))
```

Other functions relevant to various machine learning techniques are discussed in the respective chapters.

Basics of Python

Downloading and installing Python

For the purposes of this book, we will be using Python 3.5 (which is available in archived versions), the Anaconda version of which you can download from `www.continuum.io/downloads`.

Once the file is downloaded, install with all the default conditions in the installer. Once Anaconda is installed, search for "Anaconda prompt," as shown in Figure A-3.

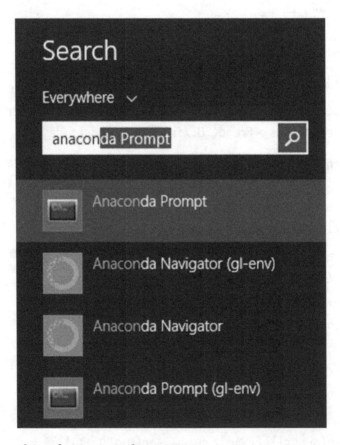

Figure A-3. *Searching for Anaconda prompt*

It takes a while for the prompt to appear (~1 minute). It looks similar to any command line or terminal program, as shown in Figure A-4.

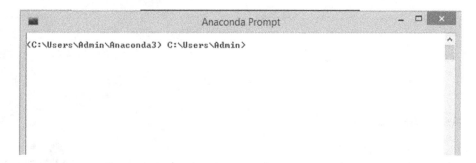

Figure A-4. *The Anaconda prompt*

Once you are able to type in the prompt, type **jupyter notebook**. A new web page opens up, as shown in Figure A-5.

Figure A-5. *The Jupyter web page*

Click the New button and then click Python 3, as shown in Figure A-6.

Figure A-6. *Selecting Python 3*

A new code editor page should appear, something like Figure A-7.

Figure A-7. *The code editor*

Type **1+1** in the space and press Shift+Enter to see if everything is working fine. It should look like Figure A-8.

Figure A-8. *The result of the addition*

Basic operations in Python

The following code shows some basic Python code (available as "Python basics.ipynb" in github).

```python
# Python can perform basic calculator type operations
1 + 1
2 * 3
1 / 2
2 ** 4
# exponential
4 % 2
# modulus operator
5 % 2
7//4
```

```
# values can be assigned to variables
name_of_var = 2
x = 2
y = 3
z = x + y
# strings can also be assigned to variables
x = 'hello'
# Lists are very similar to arrays
# They are a combination of numbers
[1,2,3]
# A list can have multiple types of data - numeric or character
# A list can also have another list
['kish',1,[1,2]]
# A list can be assigned to an object, just like a value gets assigned to a
variable
my_list = ['a','b','c']
# just like we have a word and its corresponding value in physical
dictionary
# we have a key in place of word & value in place of meaning in python
dictionary
# Dictionaries help in mapping one value to another
d = {'key1':'item1','key2':'item2'}
d['key1']
d.keys()
# A boolean is a true or false value
True
False
#Basic Python implements all of the usual operators for Boolean logic,
# but uses English words rather than symbols
# A package called "pandas" (we will work on it soon) uses & and | symbols
though for and ,
# or operations
t = True
f = False
```

```python
print(type(t)) # Prints "<type 'bool'>"
print(t and f) # Logical AND; prints "False"
print(t or f)  # Logical OR; prints "True"
print(not t)   # Logical NOT; prints "False"
print(t != f)  # Logical XOR; prints "True"
# Sets can help obtain unique values in a collection of elements
{1,2,3}
{1,2,3,1,2,1,2,3,3,3,3,2,2,2,1,1,2}
1 > 2
1 < 2
1 >= 1
1 <= 4
# Please note the usage of == instead of single =
1 == 1
'hi' == 'ahoy'
# Note how we used "and", "or"
(1 > 2) and (2 < 3)
# Writing a for loop
seq = [1,2,3,4,5]
for item in seq:
    print(item)
for i in range(5):
    print(i)
# Writing a function
def square(x):
    return x**2
out = square(2)
st = 'hello my name is Kishore'
st.split()
```

Numpy

Numpy is a fundamental package in Python which has some extremely useful functions for mathematical computations as well as abilities to work on multi dimensional data. Moreover it is very fast. We will go through a small demo of how fast numpy is when compared to traditional way of calculation, in the below code:

```
# In the below code, we are trying to sum up the square of first 10 Million
numbers
# packages can be imported as follows
import numpy as np
a=list(range(10000000))
len(a)
import time
start=time.time()
c=0
for i in range(len(a)):
    c= (c+a[i]**2)
end=time.time()
print(c)
print("Time to execute: "+str(end-start)+"seconds")
a2=np.double(np.array(a))
import time
start=time.time()
c=np.sum(np.square(a2))
end=time.time()
print(c)
print("Time to execute: "+str(end-start)+"seconds")
```

```
333333283333335000000
Time to execute using for loop: 17.9579999447seconds
3.3333328333333443e+20
Time to execute using Numpy: 0.0920000076294seconds
```

Once you implement the code, you should notice that there is a >100X improvement over traditional way of calculation using Numpy.

Number generation using Numpy

```
# notice that np automatically outputted zeroes
np.zeros(3)
# we can also create n dimensional numpy arrays
np.zeros((5,5))
# similar to zeros, we can create arrays with a value of 1
```

```
np.ones(3)
np.ones((3,3))
# not just ones or zeros, we can initialize random numbers too
np.random.randn(5)
ranarr = np.random.randint(0,50,10)
# returns the max value of array
ranarr.max()
# returns the position of max value of the array
ranarr.argmax()
ranarr.min()
ranarr.argmin()
```

Slicing and indexing

```
arr_2d = np.array(([5,10,15],[20,25,30],[35,40,45]))
#Show
arr_2d
#Indexing row
# the below selects the second row, as index starts form 0
arr_2d[1]
# Format is arr_2d[row][col] or arr_2d[row,col]

# Getting individual element value

# the below gives 2nd row first column value
arr_2d[1][0]
# Getting individual element value
# same as above
arr_2d[1,0]
# if, we need the 2nd row & only the first & 3rd column values - the below
will do the job
arr_2d[1,[0,2]]
# 2D array slicing
```

```
#Shape (2,2) from top right corner

# you can read the below as - select all rows till 2nd index & select all
columns from 1st index
arr_2d[:2,1:]
```

Pandas

Pandas is a library that helps us in generating data frames that enable us in working with tabular data. In this section, we will learn about indexing and slicing data frames and also learn about additional functions in the library.

Indexing and slicing using Pandas

```
import pandas as pd
# create a data frame
# a data frame has certain rows and columns as specified
# give the index values of the created data frame
# also, specify the column names of this data frame
df = pd.DataFrame(randn(5,4),index='A B C D E'.split(),columns='W X Y
Z'.split())
# select all the values in a column
df['W']
# select columns by specifying column names
df[['W','Z']]
# selecting certain rows in a dataframe
df.loc[['A']]
# if multiple rows and columns are to be selected - specify the index
df.loc[['A','D'],['W','Z']]
# Create a new column
df['new'] = df['W'] + df['Y']
# drop a column
# not the usage of axis=1 - which stands for doing operation at a column
level
df.drop('new',axis=1)
```

```
# we can specify the condition based on which we want to filter the data
frame
df.loc[df['X']>0]
```

Summarizing data

```
# reading a csv file into dataframe
path="D:/in-class/train.csv"
df=pd.read_csv(path)
# fetching the columns names
print(df.columns)
# if else condition on data frames is accomplished using np.where
# notice the use of == instead of single =
df['Stay_In_Current_City_Years2']=np.where(df['Stay_In_Current_City_
Years']=="4+",4, df['Stay_In_Current_City_Years'])
# specify row filtering conditions
df2=df.loc[df['Marital_Status']==0]
# get the dimension of the dataframe
df2.shape
# extract the unique values of a column
print(df2['Marital_Status'].unique())
# extract the frequency of the unique values of a column
print(df2['Marital_Status'].value_counts())
```

Index

© V Kishore Ayyadevara 2018
V. K. Ayyadevara, *Pro Machine Learning Algorithms*, https://doi.org/10.1007/978-1-4842-3564-5

W, X, Y, Z

CPSIA information can be obtained
at www.ICGtesting.com
Printed in the USA
LVHW03s1839050718
582807LV00009B/123/P